SpringerBriefs in Molecular Science

For further volumes:
http://www.springer.com/series/8898

Arnab De
Department of Immunology
Columbia University
New York, NY
USA

Rituparna Bose
Department of Earth and Environmental
    Science
The City University of New York
New York, NY
USA

Ajeet Kumar
Subho Mozumdar
Department of Chemistry
University of Delhi
Delhi
India

ISSN 2191-5407   ISSN 2191-5415 (electronic)
ISBN 978-81-322-1688-9 ISBN 978-81-322-1689-6 (eBook)
DOI 10.1007/978-81-322-1689-6
Springer New Delhi Heidelberg New York Dordrecht London

Library of Congress Control Number: 2013951797

Printed on acid-free paper

Springer is part of Springer Science+Business Media (www.springer.com)

Dedicated to
late Professor Larry Grossman,
Distinguished Professor and ex-Chairman,
Department of Biochemistry,
School of Hygiene and Public Health,
Johns Hopkins University

Preface

Pesticides have revolutionized life on this planet; however, they have also proven to be toxic for human health and environment. Indeed, the extent and severity of the toxicity was declared at the Stockholm Convention on Persistent Organic Pollutants where 9 of the 12 most dangerous and toxic organic chemicals were found to be pesticides. Traditionally, pesticides have been directed at a specific pest's life cycle as this required less pesticide and was considered to be more eco-friendly. However, this would not prevent pesticides from drifting away and potentially posing grave risks to the environment. In this book, we describe recent developments of controlled release nanoparticulate formulation of pesticides using biodegradable polymers as carrier. Technologies focusing on controlled-release of pesticide have two advantages: the pesticides are intact until sprayed and targets only the plants the pesticides are meant to protect. We have generalized the concepts to make the book useful in the post-graduate classes taught in our university and for advanced professionals alike.

While nanoparticles have revolutionized drug delivery effectively chaperoning the drug to target organs, delivery of pesticides to its intended site of action is still in the process of initial exploration. Additionally, there lies the concern of environmental safety of the fate of the pesticide-carrier. Consequently, while there have been books written on drug delivery, there are almost no available books on the topic of pesticide-delivery.

This book (a result of collaboration between scientists from Columbia University, City College of New York, and University of Delhi, India) is the first to focus exclusively on environmentally benign delivery of pesticides (controlled-release nanoparticulate formulation of pesticides using biodegradable polymers as carriers).

Acknowledgments

Some of the work embodied in this book, especially the one dealing with the role of microemulsions in synthesizing nanoparticles and using them for preparing nano-formulations of agrochemicals, has been carried out in my laboratory at the Department of Chemistry, University of Delhi. The idea of working in this field was introduced by Dr. P. K. Patanjali of IPFT, Gurgaon and his students Dr. Amit Saxena and Dr. Pooja Saxena. The funds for this work were provided by the Department of Biotechnology, Government of India and I personally thank D.B.T. and the Scientific Advisors Dr. R. R. Sinha and Dr. Dhananjay Tiwari for giving us proper encouragement for doing this work. I would also like to express my gratitude to my current Head of the Department, Professor S. M. S. Chauhan for providing us with the necessary infrastructure for setting up a laboratory where this work could be carried out. My students (both postgraduate and undergraduate) deserve a special mention, as it is their untiring effort that has resulted in this work. All of us would like to sincerely thank Mr. Sushil Mishra for helping us with all the data and also meticulously typing the entire manuscript. I also take the opportunity to thank my wife, Mrs. Jayati Mozumdar and my sons, Deepto and Aaloke, for providing me with a proper ambience and all the emotional support for writing this work. Finally, I must thank my parents Dr. Monojit Mozumdar and Dr. (Mrs.) Anjali Mozumdar for inculcating the right values in me. I decided to dedicate this book in the memory of my postdoctoral mentor, late Professor Larry Grossman. He taught me the art of doing Science and the spirit of never to give up before a goal is reached.

Subho Mozumdar

Acknowledgments

Contents

Figures

Tables

Abbreviations

°C	Degree Celsius
μg	Microgram
μm	Micrometer
Å	Angstrom
AFM	Atomic force microscope
AIBN	Azobisisobutyronitrile
ANP	Aluminium oxide nanoparticle
APG	Alkyl polyglucoside
BApNA	Nα-benzoyl-DL-arginine-p-nitroanilide hydrochloride
BmNPV	*B. mori* nuclear polyhedrosis virus
BSD	Backscattering detector
CCD	Charge-coupled device
CIBRC	Central Insecticides Board and Registration Committee
CMC	Critical micelle concentration
CRT	Cathode-ray-tube display
Cy-A	Cyclosporin A
DDT	Dichlorodiphenyltrichloroethane
DLS	Dynamic light scattering
DNA	Deoxyribonucleic acid
EC	Emulsifiable concentrate
EDX	Energy-dispersive X-ray spectroscopy
EM	Electron microscopy
EPA	US Environmental Protection Agency
ESD	Emulsification-solvent diffusion
eV	Electron volts
EW	Oil-in-water emulsions
EXTOXNET	Extension toxicology network
FAO	Food and Agriculture Organization of the United Nations
GMO	Genetically modified organisms
GR	Granular formulations
h	Hour
HCH	Gamma hexachlorocyclohexane
HGPI	Helicoverpa gut protease inhibition
HLB	Hydrophile-lipophile balance

IAEA	International Atomic Energy Agency
ICH	International Conference on Harmonization of Technical Requirements for Registration of Pharmaceuticals for Human Use
IEP	Isoelectric point
IPM	Integrated pest management
K	Kelvin
LC50	Lethal concentration 50
LFAME	Long-chain fatty acid methyl esters
MIC	Methyl isocyanate
m	Meter
ME	Microemulsion formulation
mg	Milligram
mL	Milliliter
MPa	Mega Pascal
MPEG	Methoxypolyethylene glycol
mV	Millivolts
nm	Nanometer
NMA	N-methylolacrylamide
O/W	Oil in water
O/W/O	Oil-in-Water-in-Oil
PCL	Poly(ε-caprolactone)
PCS	Photon correlation spectroscopy
PDI	Poly dipsersibility index
PEG	Polyethylene glycol
PEG-PBD	Poly(ethylene glycol)–poly(butadiene)
PHSN	Porous hollow silica nanoparticles
PI	Proteinase inhibitors
PLA	Polylactic acid
PLA-PEG	Poly(lactic acid)–poly(ethylene oxide)
PLGA	Poly(lactic-co-glycolic acid)
PMMA	Poly(methyl methacrylate)
ppm	Parts-per-millions
PVP	Polyvinylpyrrolidone
QELS	Quasi-elastic light scattering
q.s	Quantity sufficient (from Latin *quantum sufficit*)
RESS	Rapid expansion of supercritical solution
SAS	Supercritical anti-solvent
SC	Suspension concentrate
SE	Suspoemulsion formulation
SEM	Scanning electron microscopy
SFAME	Short-chain fatty acid methyl esters
SL	Solution concentrates
SLN	Solid lipid nanoparticle
SNP	Silver nanoparticles
SPM	Scanning probe microscope

TEM	Transmission electron microscopy
TI	Trypsin inhibition
TX-100	TritonX-100
UCC	Union Carbide Corporation
UCIL	Union Carbide India Limited
UNEP	United Nations Environment Programme
UNIDO	United Nations Industrial Development Organization
USA	United States of America
USEPA-OPP	U.S. Environmental Protection Agency Office of Pesticide Programs
UV	Ultra violet
VMD	Volume mean diameter
W	Watts
W/O	Water in oil
W/O/W	Water-in-oil-in-water
WP	Wettable powder

Abstract

In the current scenario the persistent challenge is to produce more food and secure the cultivated food to feed the world. The green revolution has brought tremendous increase in the worldwide crop production. Protecting the growing crop and securing the yielded gains by using pesticide has additionally helped in the production. However, the amazing performance of pesticides has encouraged their excessive use and is now causing accumulation in the environment. It has been found that the residue of pesticides can contaminate soil, water, and through crops can enter the food chain. Over the past few years there has been an increasing pressure from government and regulatory authorities to develop formulations which can have less impact on the environment and be safe for nontargeted species. In this direction, conventional formulations like granules, emulsions, and suspensions are being continuously replaced by novel formulations like micro-emulsions and multiple-emulsions and further by upcoming nano-formulations. Nano-formulations have the advantage that less quantity of pesticide can be used to target large area and thereby made to exert lesser impact on pesticide accumulation in the environment. Moreover, selection of biologically nontoxic ingredient for nano-formulations can additionally ensure the safety of the products.

Chapter 1
Introduction

The world population is increasing massively and it is said that there has been an increase from 2.5 billion in 1950 to 6.1 billion in the year 2000. This means that the population of Earth has grown more than double in the past 50 years. It is expected that by the year 2050, the world population will be about 9.1 billion. Presently, the world population is rising with an annual rate of 1.2 %, i.e., 77 million people added up per year (Carvalho 2006). India, China, Pakistan, Bangladesh, Indonesia, and Nigeria account for half of this global annual increment. However, the global population needs food every day to survive. At the worldwide level, a significant progress has been made since 1960 toward improving the nutrition and securing the food for human beings. It may be said the world gross agricultural production has grown more rapidly than the world population, with an average positive increase in the production of food per capita (Klassen 1995). However, the gap between the amount of food produced and the global population to feed is likely to increase till the year 2050. The maximum world production of cereal grains is estimated at about 3,300 metric tonnes which is 60 % more than today (Gilland 2002).

In the current scenario, the persistent challenge is to produce not only more food to feed the world but also to ensure the security of the cultivated food. It is believed that doing so will alleviate poverty and under-nourishment and this can finally result in the improvement of human health and general welfare. This aggravating global challenge for increasing demand for the food production per capita could be met by one of the several means, or a combination approach of the following:

- Increasing the land area for agriculture;
- Improving soil and water management;
- Enhancing the crop yield by using organic fertilizers;
- Controlling pests for pre-harvesting and post-harvesting damages;
- Using more productive plants and plant varieties which are resistant to the pests;
- Promoting the use of genetically modified organisms (GMOs) which are resistant to pests and diseases.

A. De et al., *Targeted Delivery of Pesticides Using Biodegradable Polymeric Nanoparticles*, SpringerBriefs in Molecular Science, DOI: 10.1007/978-81-322-1689-6_1, © The Author(s) 2014

The best and the easiest way to increase the food production is by increasing the land area of agriculture and this does not seem to be an easy task. In reality, there is an actual decrease in agricultural land (hectares per inhabitant) in all regions of the globe (Alexandratos 1999). Moreover, the rising population needs more land to live. Besides this, there are certain other reasons for loss of agricultural land which can be attributed to:

- Erosion of land;
- Decrease in the fertility of land;
- Salinization and desertification of soils.

The new land for agriculture can be only being found with the price of sacrificing green forests. This can be dangerous, and several forests have been declared or classified as ecological reserves and natural parks.

Today, the world is facing another major problem which is the scarcity of water. In continents such as Africa, Middle East, Asia (and nearly everywhere), water is becoming scarce for drinking and for irrigation as well. An average estimate for the production of 100 kg of wheat and rice requires approximately 50,000 and 200,000 L of water, respectively. In order to meet the current demands, many countries are already using the underground water for the irrigation of lands. The water pumped out from deep aquifers is estimated to be exhausted in about 20–30 years (UNEP 2004). Therefore, a better management of water resources is required and those plant varieties which are better adapted to regional weather conditions can help in increasing the water use efficiency.

However, the above solutions made to increase the crop production can suffer from their own limitations, and therefore, in the present time, the probable immediate answer can be brought by more intensive use of agrochemicals that hold promise in increasing the production of food. Agrochemicals include two large groups of compounds:

- Chemical fertilizers;
- Pesticides.

The world has already witnessed the green revolution brought by the use of chemical fertilizers, which has tremendously increased worldwide crop production since the 1960. The huge increase in production obtained from the same surface of land with the help of mineral fertilizers (based on nitrogen, phosphorus potassium) was the best result shown by the green revolution.

The green revolution brought the concept of using best yielding crops worldwide to produce a large amount of food. In this line, protecting the growing crop and securing the yielded gains by using pesticide can additionally help in the production.

The use of pesticides, including insecticides, fungicides, herbicides, rodenticides, in order to protect crops from pests, can not only allow a significant reduction to the losses but can also improve the yield of crops such as corn, maize, vegetables, potatoes, and cotton and can protect cattle from diseases and ticks. The world has known a continuous growth of pesticide usage, both in the number of chemicals and their quantities, sprayed over the fields.

Pesticides are basically poisons, which are purposely used in the environment to control pests. But at the same time, they can also act upon other species such as honeybees, birds, animals, and humans, causing serious side effects on non-target species. The residues of pesticides can contaminate soil and water as well. In addition, the residues can remain entrapped in the crops and thereafter enter the food chain. Finally, these will be ingested by humans along with food and water (Carvalho 2006). One more problem which is now aggravating is the fact that insects and pests are developing resistance to insecticides. Consequently, the dose level of pesticide for pest control needs to be increased and this will surely result in their own accumulation in the environment. One may argue that another way out is the demand for new pesticide molecules. Based on this proposition, chemical companies are continuously synthesizing new chemicals for the market.

There can be some chemical-free approaches for food production, and these can include the cultivation of better and more resistant species of crops for higher yields. Certain species of rice such as *Oriza sativa, O. japonica* have been accounted for their highest yields. Moreover, new corps can be designed by scientific approaches such as developing high-yield varieties by hybridization and genetic engineering (Khush 2002). A special role in such scientific approaches is also played by the use of radiation-induced mutations (irradiation of seeds). This technique, being promoted by FAO and IAEA, has allowed for producing healthy varieties of more productive plants (like groundnuts in India) (IAEA 2004). A yet another way of improving food security is through increasing the shelf life of foodstuff through irradiation of foods themselves (IAEA 2003). This procedure has been largely tested and demonstrated to be effective in delaying spoilage of potatoes, onions, fruits, and many other foodstuffs. Although it has the potential to replace many agrochemical additives that are not totally safe, this treatment is also not widely accepted publically (Macfarlane 2002).

The above-mentioned chemical-free approaches are either slow processes or if developed need long time for trials. The obvious choice is pesticides which can exert their own demerits. However, special care and judgment must be made for their use. The beneficial outcome from use of pesticides provides evidence that pesticides will continue to be a vital tool in the diverse range of technologies that can maintain and improve living standards for the people of the world. Some alternative methods may be more costly than conventional chemical-intensive agricultural practices, but often these comparisons fail to account for the high environmental and social costs of pesticide use. The externality problems associated with the human and environmental health effects of pesticides need to be addressed as well (Popp et al. 2013).

References

Alexandratos N (1999) World food and agriculture: outlook for the medium and longer term. Proc Natl Acad Sci 96(11):5908–5914

Carvalho FP (2006) Agriculture, pesticides, food security and food safety. Environ Sci Policy 9(7):685–692

Gilland B (2002) World population and food supply: can food production keep pace with population growth in the next half-century? Food Policy 27(1):47–63. doi:10.1016/S0306-9192(02)00002-7

IAEA (2003) Radiation processing for safe, shelf-stable and ready-to-eat food. In: IAEA TECDOC Series No. 1337, International Atomic Energy Agency, Vienna, Austria

IAEA (2004) Genetic improvement of under-utilized and neglected crops in low income food deficit countries through irradiation and related techniques. In: IAEA TECDOC Series No. 1426, International Atomic Energy Agency, Vienna, Austria

Khush GS (2002) The promise of biotechnology in addressing current nutritional problems in developing countries. Food Nutr Bull 23(4):354–357

Klassen W (1995) World food security up to 2010 and the global pesticide situation. In: Proceedings of the 8th international congress on pesticide chemistry, American Chemical Society, Washington, DC, pp 1–32

Macfarlane R (2002) Integrating the consumer interest in food safety: the role of science and other factors. Food Policy 27(1):65–80

Popp J, Pető K, Nagy J (2013) Pesticide productivity and food security: a review. Agron Sustainable Dev 33(1):243–255

UNEP (2004) Geo year book 2004/5: an overview of our changing environment. United Nations Environment Programme, Nairobi, Kenya

Chapter 2
Worldwide Pesticide Use

The worldwide consumption of pesticides is about two million tonnes per year: Out of which 45 % is used by Europe alone, 25 % is consumed in the USA, and 25 % in the rest of the world. India's share is just 3.75 %. The usage of pesticides in Korea and Japan is 6.6 and 12.0 kg/ha, respectively, whereas in India, it is only 0.5 kg/ha. Globally, the pesticides cover only 25 % of the cultivated land area. The three most commonly used pesticides are HCH (only gamma-HCH is allowed), DDT, and malathion, and these account for about 70 % of the total pesticide consumption. Despite development of newer pesticide, these pesticides still remain the choice of small farmers because they are cost-effective, easily available, and display a wide spectrum of bioactivity. Discussing the total consumption of pesticides in India (Fig. 2.1), 80 % are in the form of insecticides, 15 % are herbicides, 2 % are fungicides, and less than 3 % are others. While comparing the

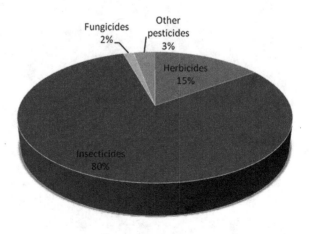

Fig. 2.1 Consumption of pesticides in the Indian scenario

A. De et al., *Targeted Delivery of Pesticides Using Biodegradable Polymeric Nanoparticles*, SpringerBriefs in Molecular Science, DOI: 10.1007/978-81-322-1689-6_2, © The Author(s) 2014

Fig. 2.2 Worldwide
consumption of pesticides

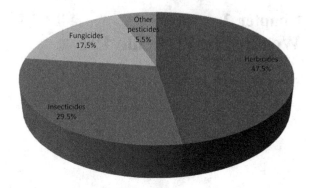

worldwide consumption of pesticide, 47.5 % is the share of herbicides, 29.5 % is the share of insecticides, 17.5 % is that of fungicides, and others account for 5.5 % only (Fig. 2.2). On the contrary, the consumption of herbicides in India is probably low, because weed control is mainly done manually by hand (Figs. 2.1 and 2.2). In addition to public health and agricultural use, pesticides also find their use in other sectors too.

Chapter 3
Pesticide Application in India

The main use of pesticides in India deals with agriculture and public health sector so as to control the numerous pests and diseases that can affect man and crop as well (Gupta 2004). Pesticide use in India began in 1948 with the import of Dichlorodiphenyltrichloroethane (DDT) for malaria control and gamma hexa-chlorocyclohexane (HCH) for locust control (Fig. 3.1a and b). HCH is also known as lindane and gammaxene.

Later in 1949, both of the pesticides, DDT and HCH, began to be used in agriculture (Joshi and Mittal 2012). The production of basic pesticide started in India in the year 1952 with the manufacture of HCH, followed by DDT. Since then, the production of pesticides has increased tremendously year by year. By 1958, India produced over 5,000 metric tonnes of pesticides, especially insecticides like DDT and HCH. In the mid-nineties, the production was approximately 85,000 metric tonnes, and about 145 pesticides were registered. Even today, the main choice of pesticide production is bulk insecticides. India has become the second largest manufacturer of basic pesticides in the Asian continent and ranks twelfth globally. In spite of such a large consumption of pesticides in India, it is estimated that crop losses due to pests vary between 10 and 30 %. In economic terms, the annual losses due to pests, despite pesticide use, amount to Rs. 290,000 million per year (Gupta 2004). By November, 2012, there was a total of 241 pesticides registered in India of which 29 pesticide molecules (including endo-sulfan) have been banned in India (CIBRC 2012).

The extensive use of pesticides has played a disastrous role with human and other life forms on the earth. Small but significant fractions of acute human poisoning have been accounted to be due to pesticides. The numbers of accidental outbreaks of poisoning by pesticides are increasing. Specially for India, the first report of pesticide poisoning came from the state of Kerala in 1958, where over 100 people died after consuming wheat flour contaminated with parathion (Fig. 3.2a) (Karunakaran 1958). The chemical used was ethyl parathion known as Folidol E 605 which was introduced by Bayer. In the same year, poisoning in Kerala caused deaths of 102 people. This was mainly due to careless handling and storage of wheat. Afterward, several cases of human and animal poisonings (besides deaths of birds and fishes) have been reported (Sethuraman 1977).

A. De et al., *Targeted Delivery of Pesticides Using Biodegradable Polymeric Nanoparticles*, SpringerBriefs in Molecular Science, DOI: 10.1007/978-81-322-1689-6_3, © The Author(s) 2014

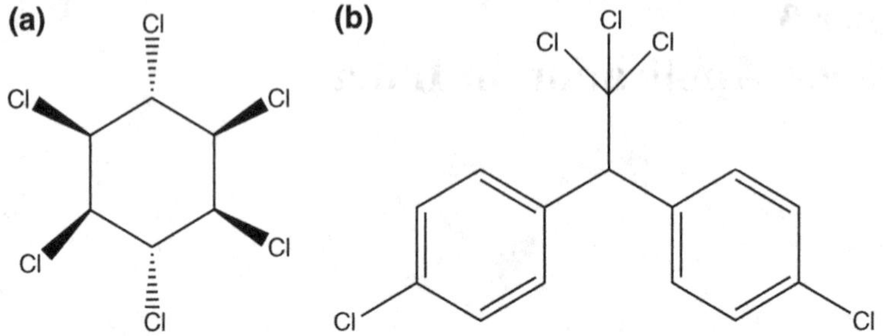

(a) **(b)**

Fig. 3.1 Chemical structure of **a** gamma hexachlorocyclohexane (HCH) and **b** dichlorodiphenyltrichloroethane (DDT)

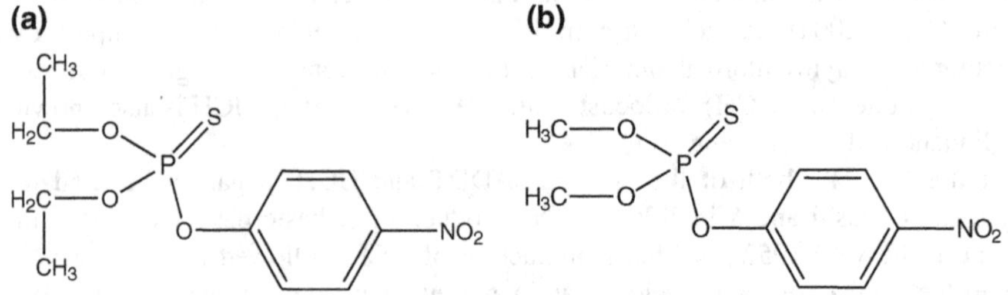

(a) **(b)**

Fig. 3.2 Chemical structure of **a** parathion and **b** malathion

In year 1967–1968, 35 cases of malathion (diazole) (Fig. 3.2b) poisoning was reported in the district Indore of Madhya Pradesh, out of which five died. ECG changes were recorded in all the cases, and furthermore, autopsy and histopathological studies revealed damage to the myocardium (Sethuraman 1977). In another report from Madhya Pradesh where 12 persons who consumed wheat for 6–12 months contaminated with aldrin dust and gammexane developed symptoms of poisoning which consisted of myoclonic jerks, generalized clonic convulsions, and weakness in the extremities (Gupta 1975). In another outbreak in 1977, eight cases of grand mal seizures were reported from a village of Uttar Pradesh following accidental ingestion of HCH-contaminated wheat (Gupta 2004).

The Bhopal gas tragedy is a catastrophe that has no parallel in the history of industries. In the early morning of December 3 1984, loud rumbling reverberated around the plant as a safety valve gave way sending a plume of methyl isocyanate (MIC) gas into the early-morning air. Within hours, the streets of Bhopal were littered with human corpses and the carcasses of buffaloes, cows, dogs, and birds. An estimated 3,800 people died immediately. Estimates of the number of people killed in the first few days by the plume from the Union Carbide India Limited (UCIL), an Indian subsidiary of Union Carbide Corporation (UCC) plant ran as high as 10,000 (with 15,000 to 20,000 premature deaths reportedly occurring in the subsequent two decades). The Indian government reported that more than half a million people were exposed to the gas. It became one of the worst chemical disasters in history, and the name Bhopal became synonymous with industrial disaster accidents (Broughton 2005).

References

Broughton E (2005) The Bhopal disaster and its aftermath: a review. Environ Health: A Global Access Sci Sour 4(1):6

CIBRC (2012) Insecticides/Pesticides Registered under section 9(3) of the Insecticides Act, 1968 for use in the Country (as on 30/11/2012). http://www.cibrc.nic.in/. Accessed 5 May 2013

Gupta P (1975) Neurotoxicity of chronic chlorinated hydrocarbon insecticide poisoning-a clinical and electroencephalographic study in man. Indian J Med Res 63(4):601

Gupta P (2004) Pesticide exposure—Indian scene. Toxicology 198(1):83–90

Joshi TK, Mittal A (2012) Need for a coherent pesticide policy in India intergovernmental forum on chemical safety. Forum Standing Committee, WHO. http://www.who.int/ifcs/FSC/forumsc/cdPesticides/Documents/Joshi_Need_for_Coherent_Pesticide_Policy_India.pdf. Accessed 3 Apr 2013

Karunakaran C (1958) The Kerala food poisoning. J Indian Med Assoc 31(5):204

Sethuraman V (1977) A case of BHC poisoning in a heifer calf [dairy cattle, India]. Indian Veterinary J 54:486–487

Chapter 4
Food Contamination and Wastage by Insects

Insects are found in all types of environment and occupy a little more than two-thirds of the known species of animals in the world. Insects can infest all kinds of plants (including crop plants, forest trees, medicinal plants, and weeds). Moreover, they can feed on the food and other stored products in warehouses, bins, storage structures and packages, causing huge amount of loss to the stored food and also deterioration of the food quality. Insects can inflict injuries to plants and stored products either directly or indirectly in their attempts to secure food. The insects that cause 5–10 % damages are called minor pests and those that cause damages above 10 % are considered as major pests (Navarajan Paul 2007; Dhaliwal et al. 2010). Insects that cause injury to plants and stored products are grouped into two major groups, namely chewing insects and sucking insects. The former group chews off plant parts and swallow them, thereby causing damage to the crops. Sucking insects pierce through the epidermis and suck the sap. Many of the sucking insects serve as vectors of plant diseases and also inject their salivary secretions containing toxins that cause severe damage to the crop (Navarajan Paul 2007).

Pimentel (2009) reviewed that worldwide loss caused by insect pests has been estimated at about 14 % and that by plant pathogens about 13 % and weeds at about 13 %. The value of this crop loss was estimated to be US $2,000 billion per year.

Traditional strategies such as crop rotation, healthy crop variety, manipulations in sowing dates, integrated pest management (IPM) have been commonly used by farmers for the management of insect pest in agriculture. Among these, IPM is the most popular approach. The term "IPM" was formalized by the US Academy of Sciences in 1969. IPM was introduced as a solution to avoid the side effects of pesticide, which combines the use of different pest control strategies (cultural, resistant varieties, biological and chemical control). IPM is thus more complex for the producer to implement, as it requires skill in pest monitoring and in understanding the pest dynamics (besides the cooperation of all among the producers for effective implementation) (Rai and Ingle 2012).

In the 1960s when the IPM began to be promoted as a pest control strategy, there were only a few IPM technologies available for field application. In the

A. De et al., *Targeted Delivery of Pesticides Using Biodegradable Polymeric Nanoparticles*, SpringerBriefs in Molecular Science, DOI: 10.1007/978-81-322-1689-6_4, © The Author(s) 2014

1970s, extensive research on the management of insect pest generated some novel products and knowledge for successful implementation of IPM in crops such as rice, cotton, sugarcane, and vegetables. However, the exaggerated expectations about the possibility that dramatic reduction in pesticide use could be achieved without significant decline in crop yields as a result of adoption of IPM could not be realized (Rai and Ingle 2012).

IPM is an ecologically based strategy that focuses on long-term solution of the pests through a combination of techniques such as biological control, habitat manipulation, modification of agronomic practices, and the use of resistant varieties. Embracing a single method to control a specific organism does not constitute IPM, even if the procedure is an essential element of the IPM system. Integration of multiple pest suppression techniques has the highest probability of sustaining long-term crop protection. Pesticides may be used to remove/prevent the target organism, but only when assessment with the help of monitoring and scouting methods indicate that they are needed to prevent economic damage. Pest control strategies, including pesticides, should be carefully selected and applied so as to minimize risks to the human health, non-target organisms, and environment (Rai and Ingle 2012).

In the context of crop protection, sustainability refers to the substitution of chemicals and capital with farm-grown biological inputs and knowledge, aimed at reduction in the cost of production without lowering the yields. Sustainability builds on the current agricultural achievements, adopting a sophisticated approach that can maintain high yields and farm profits without degrading the resources. Sustainable agriculture is a reality based on the human goals and on the understanding of the long-term impacts of human activities on the environment and on other species. This philosophy combines the application of prior experience and the latest scientific advancements so as to create an integrated, resource-conserving and equitable farming system. The systems approach can minimize environmental degradation, sustain agricultural productivity, promote economic viability (in both the short and long run), and maintain the quality of the life (Rai and Ingle 2012). Sustainable farming practices commonly include:

- Crop rotations that can mitigate weeds, disease, insect, and other pest problems provide alternative sources of soil nitrogen, and reduce soil erosion and risk of water contamination by agricultural chemicals.
- Pest control strategies, including IPM techniques, that can reduce the need for pesticides by practices such as scouting/monitoring, use of resistant cultivars, timing of planting, and biological pest controls.
- Increased mechanical/biological weed control, more soil and water conservation practices, and strategic use of green manures.
- Use of natural or synthetic inputs in a way that poses no significant hazard to humans or the environment.

References

Dhaliwal G, Jindal V, Dhawan A (2010) Insect pest problems and crop losses: changing trends. Indian J Ecol 37(1):1–7

Navarajan Paul AV (2007) Insect pests and their management. Indian Agricultural Research Institute, New Delhi

Pimentel D (2009) Pesticides and pest control. In: Peshin R, Dhawan A (eds) Integrated pest management: innovation-development process. Springer Netherlands, pp 83–87. doi:10.1007/978-1-4020-8992-3_3

Rai M, Ingle A (2012) Role of nanotechnology in agriculture with special reference to management of insect pests. Appl Microbiol Biotechnol 94(2):287–293

References

Chapter 5
Pesticide Formulations

Pesticide formulation is the process of transforming a pesticidal chemical into a product, which can be applied by practical methods to permit its effective, safe, and economical use. A pesticide formulation is an active chemical with inert ingredients, which can provide effective and economic control of pests (UNIDO 1983).

The primary objectives of formulation technology are to optimize the biological activity of the pesticide and to give a product, which is safe and convenient for use. However, because of the wide variety of pesticides that are available, many different types of formulations have been developed depending mainly on the physicochemical properties of the active ingredients (Knowles 2008).

5.1 Conventional Formulations

Formulations based on the older technologies are still available and used. They represent the greatest volume of products applied to crops. A brief review is given here of the main types of conventional formulations.

5.1.1 Granules

Granular (GR) formulations (UNIDO 1983) are used for direct broadcasting to the field often as preemergence herbicides or as soil insecticides. The active ingredient concentration is usually from 1 to 40 %, and the granule mesh size is generally between 250 and 1,000 μ. The granules should be non-caking, non-dusty, free flowing and should disintegrate in the soil so as to release the active ingredient. Granules are usually made either by coating a fine powder onto a substrate, e.g., sand (using a sticker such as PVP solution), or by solvent impregnation onto an absorbent carrier. Resins or polymers may be sprayed onto the granules to control release rates (Knowles 2008). GR carriers can be broadly categorized into mineral

A. De et al., *Targeted Delivery of Pesticides Using Biodegradable Polymeric Nanoparticles*, SpringerBriefs in Molecular Science, DOI: 10.1007/978-81-322-1689-6_5, © The Author(s) 2014

and organic types. The mineral carriers include sand, limestone, gypsum, kaolin, montmorillonite, attapulgite, and diatomite. The organic GR carriers include corncobs, pecan shells, peanut hulls, and recycled paper fiber (Collins et al. 1996). The absorptive capacity of the carrier is an important parameter and is a function of the crystalline structure and the available surface area of the carrier particles. Granules are becoming less popular now because of the increasing use of post-emergence herbicides.

5.1.2 Solution Concentrates

Solution concentrates (SLs) can be the simplest formulation to make. It is an aqueous solution of the active ingredient, which merely requires dilution in a spray tank. The number of pesticides that can be formulated in this way is limited by water solubility and hydrolytic stability of the active ingredient. Some solution concentrate formulations contain a surfactant (usually a nonionic ethylene oxide condensate) to assist wetting onto the leaf surface (Knowles 2008). Solution concentrate formulations are usually very stable and, therefore, have few storage problems. Some problems do occur occasionally, such as precipitation during dilution and corrosion of metal containers or spray applicators. However, these problems can be overcome by the use of suitable additives, such as cosolvents and corrosion inhibitors. The composition of a typical solution concentrate formulation is shown in Table 5.1.

Table 5.1 Approximate quantity of the excipients used to formulate various pesticide formulations

Excipients	Solution concentrate Quantity (%w/w)	Emulsion concentrate Quantity (%w/w)	Wettable powders Quantity (%w/w)	Suspension concentrate Quantity (%w/w)
Active ingredient	20–50	20–70	25–80	20–50
Antifreeze agent	5–10	–	–	5–10
Dispersing agent	–	–	2–5	2–5
Wetting agent	3–10	–	1–3	2–5
Anti-settling agent	–	–	–	0.2–2
Emulsifying agent	–	5–10	–	–
Solvents	–	q.s. 100	–	–
Inert filler	–	–	q.s. 100	–
Water	q.s. 100	–	–	q.s. 100

Nonylphenol or tallow amine ethoxylates are often used as tank mix wetters for solution concentrate formulations to enhance bioefficacy. Alternatively, the wetting agent may be built into the formulation to ensure that the correct rate of wetting agent is applied so as to optimize biological activity. This is often the case, for example, with paraquat and glyphosate formulations. A considerable amount of work is being carried out on new surfactant wetting agents for glyphosate formulations (Knowles 2008). In some cases, preservatives may be necessary to prevent mold growth or bacterial spoilage during long-term storage.

5.1.3 Emulsifiable Concentrates

Emulsifiable concentrate (EC) formulations have been very popular for many years and represent the biggest volume of all pesticide formulations in terms of consumption worldwide. ECs are made from oily active ingredients or from low melting, waxy solid active ingredients, which are soluble in nonpolar hydrocarbon solvents (such as xylene, C9–C10 solvents, solvent naphtha, odorless kerosene, or other proprietary hydrocarbon solvents). Surfactant emulsifiers are added to these formulations to ensure spontaneous emulsification with good emulsion stability properties in the spray tank. Careful selection of a "balanced pair" emulsifier blend is frequently necessary to ensure that emulsion dilution stability is maintained over widely differing climatic conditions and degrees of water hardness. Emulsion droplets of $0.1–5~\mu$ are produced when the formulation is mixed with water. The formulation of ECs has been greatly facilitated by the commercial development, over the last 20 years, of nonionic emulsifying agents in which the hydrophilic portion of the molecule consists of a polyethylene oxide chain. The nonionic surfactant that is commonly used is a nonylphenol hydrophobic chain condensed with 12 or more moles of ethylene oxide (Knowles 2008). The other component of the balanced pair is generally an anionic surfactant such as the oil-soluble calcium salt of dodecylbenzene sulfonic acid. However, the nonylphenol ethoxylates have been suspected of having endocrine-modulating properties. It is because of this potentially toxic effect, alternative ethylene oxide condensates based on aliphatic alcohol hydrophobes are being introduced. The total concentration of the emulsifier blend is usually 5–10 % of the formulation. There are no definite rules to determine the ratio of anionic to nonionic surfactant in the mixed emulsifiers, but guidance can be obtained from the HLB system. HLB stands for hydrophile–lipophile balance, and a general rule of the thumb is the higher the HLB value the more hydrophilic (water-soluble) is the surfactant. The HLB range of 8–18 normally provides a good oil-in-water emulsion. The optimum ratio of anionic and nonionic surfactants is determined experimentally to give spontaneous emulsification in water and to give a stable emulsion with very little creaming and no oil droplet coalescence. ECs are limited in the number of active ingredients for which they are suitable. Many pesticides are not soluble enough to be supplied economically in this form. However, it may be possible to boost the solubility of the active ingredient by the addition of a more polar solvent without increasing the risk of crystallization in the spray tank. The composition of a typical EC formulation is shown in Table 5.1.

The presence of solvents and emulsifiers in emulsion concentrate formulations can sometimes give enhanced biological efficacy compared with other formulations. Many insecticides, e.g., organophosphorous compounds and pyrethroids, are oil-soluble liquids or waxy solids and are readily formulated as ECs, and a few active ingredients need to be formulated with solvents for optimum biological activity.

5.1.4 Wettable Powders

Wettable powder (WP) formulations of pesticides have been known for many years and are made usually from solid active ingredients with high melting points, which are suitable for dry grinding through a mechanical grinder such as a hammer- or pin-type mill or by air milling with a fluid energy micronizer. Air milling gives much finer particles ($5-10$ μ) than mechanical milling ($20-40$ μ) and can also be more suitable for active ingredients with lower melting points. However, care must be taken to prevent suppress or contain dust explosions, which may occur if a source of ignition (such as static energy) is present in both types of mills. WPs usually contain dry surfactants as powder wetting and dispersing agents and inert carriers or fillers. They frequently contain more than 50 % active ingredient, and the upper limit is usually determined by the amount of inert material such as silica required to prevent the active ingredient particles from fusing together during processing in the dry grinding mills (Knowles 2008). An inert filler such as kaolin or talc may also be needed to prevent the formulated product from caking or aggregating during storage. The excipients used to prepare WP formulation are shown in Table 5.1.

5.1.5 Suspension Concentrates

Suspension concentrate (SC) technology has been increasingly applied to the formulation of many solid crystalline pesticides since the early 1970s. Pesticide particles may be suspended in an oil phase, but it is much more appropriate for SCs to be dispersions in water. Considerable attention has been given in recent years to the production of aqueous SCs by high-energy wet grinding processes such as bead milling. The use of surfactants as wetting and dispersing agents has also led to a great deal of research on the colloidal and surface chemistry aspects of dispersion and stabilization of solid–liquid dispersions (Knowles 2008). Water-based SC formulations offer many advantages such as

- high concentration of insoluble active ingredients
- ease of handling and application
- safety to the operator and environment
- relatively low cost
- enable water-soluble adjuvants to be built-in for enhanced biological activity.

Farmers generally prefer SCs to WPs because they are non-dusty and easy to measure and pour into the spray tank. However, there are some disadvantages, notably the need to produce formulations which do not separate badly on storage and also a need to protect the product from freezing which may cause aggregation of the particles. In most cases, SCs are made by dispersing the active ingredient powder in an aqueous solution of a wetting and dispersing agent using a high shear mixer to give a concentrated premix, followed by a wet grinding process in a bead mill to give a particle size distribution in the range 1–10 μ. The wetting/dispersing agent aids the wetting of the powder into water and breaking of the aggregates, agglomerates, and single crystals into smaller particles. In addition, the surfactant that becomes adsorbed onto the freshly formed particle surface during the grinding process prevents reaggregation of the small particles and can ensure colloidal stability of the dispersion. Typical wetting/dispersing agents used in SC formulations are as follows:

- sodium lignosulfonates
- sodium naphthalene sulfonate–formaldehyde condensates
- aliphatic alcohol ethoxylates
- tristyrylphenol ethoxylates and esters
- ethylene oxide/propylene oxide block copolymers.

More recently available are polymeric surfactants, such as "comb" surfactants, which can adsorb strongly on particle surfaces and can give considerably improved stabilization of SCs for long-term storage. A typical SC formulation is shown in Table 5.1.

The anti-settling agent is added to increase the viscosity and build up a three-dimensional network structure to prevent separation of particles during long-term storage. The anti-settling agent is usually a swelling clay such as bentonite (sodium montmorillonite) and may be mixed with water-soluble polymers to give synergistic rheological effects. The water-soluble polymers are often cellulose derivatives, natural gums, or other types of polysaccharides, such as xanthan gum, and they are generally susceptible to microbial attack. It is for this reason, preservatives are usually added to SC formulations so as to prevent degradation of the anti-settling agent and to ensure that the long-term stability of the product is not impaired. A great deal of research has been carried out using rheological techniques so as to measure the forces acting between particles and polymers and in order to enable storage stability to be predicted. However, it is still necessary to carry out long-term storage tests over a range of temperatures so as to ensure that the particles do not aggregate or separate irreversibly under normal storage conditions in the sales pack.

5.2 New Generation Formulations

Over the past few years, there has been increasing pressure from government and regulatory authorities to develop formulations, which have less impact on the environment generally. The main issues that are being addressed are as follows:

- safety in manufacture and use
- convenience for the user
- ease of pack disposal (or reuse)
- reduction in the amount of pesticide applied
- reduction in waste and effluent of all kinds.

Hence, the current trends in the development of pesticide formulations are to

- eliminate solvents wherever possible and use aqueous emulsions or microemulsions (MEs)
- replace dusty powders by SCs or water-dispersible granules
- develop multiple active ingredient formulations where appropriate
- build into the formulation bioenhancing surfactant adjuvants
- control release rate and targeting of pesticides by encapsulation techniques or seed treatment applications
- develop novel formulations such as tablets or gels
- develop more effective spray tank adjuvants to enhance biological activity and reduce pesticide dosage.

These complex requirements are being met by technical advances in surfactants and other formulation additives (particularly blends of surfactants) together with developing more powerful dispersing agents and obtaining a better understanding of the principles of colloid/surface chemistry and rheology. The ideal product would seem to be one which is free from volatile solvents, gives very low operator exposure hazard, has the maximum biological activity at the lowest dose level, and produces the minimum of pack disposal problems. Water-dispersible granules or WPs in water-soluble sachets, which can be added directly to a spray tank, can go a long way toward meeting these requirements. Development work is being carried out on these options by all the major agrochemical companies. However, it is to be noted that it will never be possible to formulate all the active ingredients this way and so other options are being evaluated extensively, along with ideas for more convenient packaging and closed transfer spray tank application systems (Knowles 2008). Aqueous-based formulations are regarded as safe alternatives to water-dispersible granule formulations and these options include (in addition to SCs which have already been discussed):

- suspoemulsions (SEs)
- o/w emulsions (concentrated emulsions)

- microemulsions
- multiple emulsions
- microcapsules (capsule suspensions).

5.2.1 Suspoemulsions

Mixed combination formulations are becoming more popular because of their convenience. It is to be ensured that the farmer applies the correct amount of each component pesticide and overcomes problems of tank mix incompatibility (Knowles 2008). If one active ingredient is a solid and the other is a liquid, it is necessary to produce a suspoemulsion formulation, which can consist of three phases:

- Solid dispersed particles
- Liquid oil droplets
- Continuous phase (usually water).

SE can, therefore, be considered to be mixtures of SCs and oil-in-water emulsions (EWs) with added surfactants (to prevent flocculation) and thickeners (to prevent separation of the dispersed phases). Surfactants used as dispersing agents for the solid phase can be similar to those already mentioned for SCs. Emulsifiers for the oily liquid phase can be similar to those used for EW. As these formulations are aqueous based and generally thickened with polysaccharides, it is necessary to add a preservative to prevent degradation of the thickener. Careful selection of the appropriate dispersing and emulsifying agents is necessary to overcome the problem of heteroflocculation between the solid particles and the oil droplets. Hence, extensive storage testing of these formulations is necessary.

5.2.2 Oil-in-Water Emulsions

EWs are now receiving considerable attention because of the need to reduce or eliminate volatile organic solvents for safer handling. It can be because they are water based, EW can have significant advantages over ECs in terms of cost and safety in manufacture, transportation, and use. However, they require careful selection of surfactant emulsifiers to prevent flocculation, creaming, and coalescence of the oil droplets (Knowles 2008). Nonionic surfactants, block copolymers, and other polymeric surfactants are now being used to produce stable emulsions. In the case of nonionic surfactants, it is sometimes useful to combine a low and a high HLB surfactant to give an emulsifier mixture with an average HLB of 11–16 for optimum emulsion stability (Knowles 2008). Droplet size is also a good indicator of stability and should be below 2 μ [volume mean diameter (VMD)]. The emulsions are usually thickened with polysaccharides such as xanthan gum to prevent separation of the oil droplets. Sometimes polymers such as polyvinyl alcohol are used as both emulsifier and thickener/stabilizer.

5.2.3 Microemulsions

MEs are thermodynamically stable transparent emulsion systems that are stable over a wide range of temperature. They have a very fine droplet size of less than 0.05 μ (50 nm) and consist of three components, namely

- oily liquid or solid dissolved in organic solvent
- water
- surfactant/cosurfactant system.

These components form a single phase containing relatively large "swollen micelles" in which the non-aqueous phase of the active ingredient and solvent are dissolved or solubilized by the surfactant system. In the preparation of micro-emulsions, two different types of surfactants are needed: one water soluble and one oil soluble. The water-soluble surfactant is usually anionic or nonionic with a very high HLB value, and the hydrophobic part of the molecule should match the oil. The cosurfactant should be oil soluble and should have a very low HLB value (for example hexanol). The total concentration of surfactants for a microemulsion can be as high as 10–30 % or more (compared to about 5 % for a typical O/W emulsion). Microemulsions have relatively low active ingredient concentrations, but the high surfactant content and solubilization of the active ingredient can give rise to enhanced biological activity (Knowles 2008).

5.2.4 Multiple Emulsions

Multiple emulsions are another class of emulsions, which can be water-in-oil-in-water (W/O/W) or oil-in-water-in-oil (O/W/O). These are complex formulations which require very careful selection of surfactant emulsifiers and stabilizers to overcome physical instability problems. Multiple emulsions are still in the research phase and could be of interest to reduce the toxicity of an active ingre-dient by restricting it to the primary internal emulsion droplet phase (Knowles 2008). However, because of the need to form a second emulsion phase, the final product must be of low active ingredient content.

References

Collins HM, Hall FR, Hopkinson M, Pesticides ACE-o (1996) Pesticide formulations and application systems: 15th volume, vol no. 1268. ASTM

Knowles A (2008) Recent developments of safer formulations of agrochemicals. Environmentalist 28(1):35–44

UNIDO (1983) Formulation of pesticides in developing countries. http://pdf.usaid.gov/pdf_docs/PNAAQ247.pdf. Accessed 16 May 2013

Chapter 6
Trends and Limitations in Chemical-Based Pest Management

Until the beginning of the twentieth century, farmers relied exclusively on cultural practices such as crop rotation, healthy crop variety, manipulations in sowing dates, etc. to manage the insect pests. Although use of pesticides began in the 1870s with the development of arsenical and copper-based insecticides and most of the pesticides were originally based on the toxic heavy metals such as arsenic, mercury, lead, and copper (Davies et al. 2007), it was the discovery of dichloro-diphenyltrichloroacetic acid (DDT), having pesticidal properties during the World War II, that revolutionized the pest control. DDT was effective at low concentration against almost all insect species. In addition, it was supposed to be less expensive and harmless to human beings, animals, and plants (Davies et al. 2007). Therefore, farmers were amazed with its effectiveness and started to use it increasingly (particularly during the green revolution era). As a result of increasing demand, the pesticide industry expanded rapidly, leading to research toward development of synthetic organic insecticides and other chemicals so as to control the pests. The negative effects of chemical pesticides, however, started emerging soon after the introduction of DDT. Producers then turned to much more toxic organophosphates and pyrethroid insecticides, and this resulted in the development of resistant strains.

It has been found that pesticides often kill the natural enemies along with the pests. With natural enemies eliminated, it became often difficult to prevent recovered pest populations from exploding to higher and more damaging levels, leading to often development of resistance to chemical pesticides. Initially, the benefits from pest control were not huge due to use in low amount. But very soon DDT became popular, and its use was increased enormously, which undoubtedly resulted in the increase in yields, but on the other hand, their adverse effects on the environment and human health also soon became apparent.

Indiscriminate, excessive, and continuous use of pesticides created a powerful selection pressure for altering the genetic makeup of the pests. Naturally resistant individuals in a pest population were able to survive onslaughts of the pesticides, and the survivors could pass on the resistance traits to their subsequent generations. This resulted in a much higher percentage pest population being resistant to pesticides (Biyela et al. 2004 and Levy 2002). The number of weed species

A. De et al., *Targeted Delivery of Pesticides Using Biodegradable Polymeric Nanoparticles*, SpringerBriefs in Molecular Science, DOI: 10.1007/978-81-322-1689-6_6, © The Author(s) 2014

resistant to herbicides was estimated to be 270, and plant pathogens resistant to fungicides were 150. Resistance to insecticides also become common, and more than 500 insect species acquired resistance to the pesticides. Due to these limitations of chemical pesticides and their hazardous effects on human beings, animals, and on fields (like loss of fertility due to its excess use and killing of beneficial soil microflora), researchers turned toward the direction of new potential agents (having minimum or no side effects) against insect pest.

References

Biyela P, Lin J, Bezuidenhout C (2004) The role of aquatic ecosystems as reservoirs of antibiotic resistant bacteria and antibiotic resistance genes. Water Sci Technol: J Int Assoc Water Pollut Res 50(1):45

Davies T, Field L, Usherwood P, Williamson M (2007) DDT, pyrethrins, pyrethroids and insect sodium channels. IUBMB Life 59(3):151–162

Levy SB (2002) Factors impacting on the problem of antibiotic resistance. J Antimicrob Chemother 49(1):25–30

Chapter 7
Biological Control of Insect Pests

Natural enemies of insect pests play a key role in reducing the levels of pest populations below those causing economic injury. Both natural and applied biological control tactics can be important in successful management of pest populations. Biological control has the advantage of being self-perpetuating (once established), and it usually does not harm non-target organisms found in the environment. In addition, it is non-polluting or (as disruptive to the environment) as chemical pesticides, nor does it leave residues on food (a concern of many people today). However, the use of biological control does require detailed knowledge of the pest's biology and the natural enemies associated with the pest and their impact.

Many biological agents have been used for the biocontrol of insect pests. However, only bacteria and fungi have been found to be most important. Bacteria used for biological control usually infect insects via their digestive tracts. *Bacillus thuringiensis* is the most widely applied species of bacteria used for biological control of *lepidopteran* (moth, butterfly), *coleopteran* (beetle), and *dipteran* (true flies) (Frederick and Caesar 1999).

Fungi that cause disease or infection in insects are known as entomopathogenic fungi, and these include at least 14 species of entomophthoraceous fungi, which can attack aphids. Species of the genus *Trichoderma* are used to manage some soilborne plant pathogens. *Beauveria bassiana* is used to manage different types of pests such as whiteflies, thrips, aphids, and weevils (Thungrabeab and Tongma 2007). Some examples of entomopathogenic fungi are as follows:

- *B. bassiana*—used against whiteflies, thrips, aphids, and weevils;
- *Paecilomyces fumosoroseus*—used against whiteflies, thrips, and aphids;
- *Metarhizium* sp.—used against beetles, locusts, Hemiptera, spider mites, and other pests;
- *Lecanicillium lecanii*—used against whiteflies, thrips, and aphids; and
- *Cordyceps* species—used against wide spectrum of arthropods.

A. De et al., *Targeted Delivery of Pesticides Using Biodegradable Polymeric Nanoparticles*, SpringerBriefs in Molecular Science, DOI: 10.1007/978-81-322-1689-6_7, © The Author(s) 2014

References

Frederick B, Caesar A (1999) Analysis of bacterial communities associated with insect biological control agents using molecular techniques. In: Proceedings of the X international symposium on biological control of weeds, pp 4–14

Thungrabeab M, Tongma S (2007) Effect of entomopathogenic fungi, Beauveria bassiana (Balsam) and Metarhizium anisopliae (Metsch) on non target insects. Kmitl Sci Technol J 7(1):8–12

Chapter 8
Management of Insect Pests Using Nanotechnology: As Modern Approaches

The above mentioned limitations and less efficiency of traditional methods have led to the development of new and modern approaches for management of insect pest, and this has become the need of the hour. Keeping in the mind the applications of nanotechnology in agriculture, it can be suggested that the use of nanomaterials will result in the development of efficient and potential approaches toward the management of insect pest. However, the literature available on this topic brings to a close conclusion that only a few researchers all over the world are working in this area, and hence, there is a pressing need to apply nanotechnology and this warrants detailed study in this field. Keeping this idea in mind, the research studies carried out (related to management of insect pest) have been reviewed here.

Previous studies have confirmed that metal nanoparticles can be effective against plant pathogens, insects, and pests. Hence, nanoparticles can be used in the preparation of new formulations such as pesticides, insecticides, and insect repellants (Barik et al. 2008; Gajbhiye et al. 2009; Goswami et al. 2010; Owolade and Ogunleti 2008). Nanotechnology has promising applications in nanoparticle-mediated gene (DNA) transfer. It can be used to deliver DNA and other desired chemicals into plant tissues for protection of the host plants against insect pests (Rai and Ingle 2012).

Porous hollow silica nanoparticles (PHSNs) loaded with validamycin (pesticide) can be used as efficient controlled release delivery system for water-soluble pesticide. Such controlled release behavior of PHSNs makes them promising carriers in agriculture (especially for controlled delivery of pesticides whose immediate as well as prolonged release can be essential for plants) (Liu et al. 2006). According to Wang et al. (2007), oil in water (nanoemulsions) can be useful for the formulations of pesticides and these could be effective against various insect pests in agriculture. Similarly, essential oil-loaded solid lipid nanoparticles can also be useful for the formulations of nanopesticides (Liu et al. 2006).

Nanosilica, a type of unique nanomaterial, is prepared from silica. It has many applications in medicine and drug development. Recently, it has been found to be useful as a catalyst and most importantly has been found to be useful as nanopesticide. Barik et al. (2008) have reviewed the use of nanosilica as nanopesticide.

A. De et al., *Targeted Delivery of Pesticides Using Biodegradable Polymeric Nanoparticles*, SpringerBriefs in Molecular Science, DOI: 10.1007/978-81-322-1689-6_8, © The Author(s) 2014

The mechanism of control of insect pest using nanosilica is based on the fact that insect pests use a variety of cuticular lipids for protecting their water barrier and thereby prevent death from desiccation. Typically, nanosilica gets absorbed into the cuticular lipids by physisorption and thereby (when applied on leaves and stem surface) causes death of insects purely by physical means. Surface charged, modified, hydrophobic nanosilica (~ 3–5 nm) can be successfully used to control a range of agricultural insect pests and animal ectoparasites of veterinary importance (Ulrichs et al. 2005).

Yang et al. (2009) have demonstrated the insecticidal activity of polyethylene glycol-coated nanoparticles loaded with garlic essential oil against adult *Tribolium castaneum* found in stored products. It has been observed that the control efficacy against adult *T. castaneum* was about 80 %, presumably due to the slow and persistent release of the active components from the nanoparticles.

Goswami et al. (2010) have studied the applications of different kinds of nanoparticles, viz. silver (SNP), aluminum oxide (ANP), zinc oxide, and titanium dioxide nanoparticles in the control of rice weevil and grasserie disease in silkworm (*Bombyx mori*) caused by *Sitophilus oryzae* and baculovirus BmNPV (*B. mori* nuclear polyhedrosis virus), respectively. In their study, they performed bioassay, in which they prepared solid and liquid formulations of the abovementioned nanoparticles; later, they applied these formulations on rice, kept them in a plastic box with 20 adults of *S. oryzae* and observed the effects for 7 days. It was reported that hydrophilic SNP was most effective on the first day. On day 2, more than 90 % mortality was obtained with SNP and ANP. After 7 days of exposure, 95 % mortality and 86 % mortality were reported with hydrophilic and hydrophobic SNP and nearly 70 % of the insects were killed when the rice was treated with lipophilic SNP. However, 100 % mortality was observed in case of ANP. Similarly, in another bioassay carried for grasserie disease in silkworm (*B. mori*), a significant decrease in viral load was reported when leaves were treated with an ethanolic suspension of hydrophobic aluminosilicate nanoparticles.

Bhattacharyya et al. (2010) have reviewed that nanotechnology will revolutionize agriculture including pest management in the near future. It is also forecasted that over the next two decades, the "green revolution" would be accelerated by means of nanotechnology. One of the examples of this technology is nanoencapsulation. It is currently used as the most important and promising approach for protection of host plants against insect pests. Nanoencapsulation includes the use of a different kind of nanoparticles with insecticide inside. In this process, a chemical such as an insecticide is slowly but efficiently released to a particular host plant for insect pest control. Nanoencapsulation with nanoparticles can allow for proper absorption of the chemical into the plants unlike the case of larger particles (Scrinis and Lyons 2007).

Stadler et al. (2010) for the first time studied the insecticidal activity of nanostructured alumina against two insect pests, viz. *S. oryzae* (L.) and *Rhyzopertha dominica* (F.), which are major insect pests in stored food supplies throughout the world. They reported significant mortality after 3 days of continuous exposure to nanostructured alumina-treated wheat. Therefore, as compared to commercially available insecticides, inorganic nanostructured alumina may provide a cheap and reliable alternative for control of insect pests and such studies may expand the frontiers for nanoparticle-based technologies in pest management.

8.1 Additional Applications of Nanotechnology in the Field of Agriculture

Nanotechnology has the potential to revolutionize different sectors of the agricultural and food industry with modern tools for the treatment for diseases, rapid disease detection, enhancing the ability of plants to absorb nutrients, etc. Smart sensors and smart delivery systems will help the agricultural industry combat viruses and other crop pathogens (Rickman et al. 2003). Nanotechnology will also protect the environment indirectly through the use of alternative (renewable) energy supplies and filters or catalysts to reduce pollution and clean up existing pollutants (Tungittiplakorn et al. 2004).

8.1.1 Precision Farming

Precision farming has been a long-desired goal to maximize output (i.e., crop yields) while minimizing input (i.e., fertilizers, pesticides, herbicides) through monitoring environmental variables and applying targeted action. Precision farming makes use of computers, global satellite positioning systems, and remote sensing devices to measure highly localized environmental conditions, thus determining whether crops are growing at maximum efficiency or precisely identifying the nature and location of problems. Precision farming can also help in reducing agricultural waste and thus keep environmental pollution to a minimum. Although not fully implemented yet, tiny sensors and monitoring systems enabled by nanotechnology will have a large impact on future precision farming methodologies. Ultimately, precision farming, with the help of smart sensors, will allow enhanced productivity in agriculture by providing accurate information, thus helping farmers to make better decisions (Joseph and Morrison 2006).

8.2 Nanomaterials: Antimicrobial Agents for Plant Pathogens

Antimicrobial activity of different metal nanoparticles, particularly copper and silver nanoparticles, has been investigated by some researchers against the plant pathogens. Cioffi et al. (2004) have reported the antifungal activity of polymer-based copper nanocomposites against plant pathogenic fungi. Park et al. (2006) have studied the efficacy of nanosized silica–silver (silica–silver nanoparticles) in the control of plant pathogenic fungi, viz. *Botrytis cinerea*, *Rhizoctonia solani*, *Colletotrichum gloeosporioides*, *Magnaporthe grisea*, and *Pythium ultimum*. They also demonstrated the effect of nanobased products prepared from these nano-particles against the powdery mildew disease of pumpkin and found that the disease-causing pathogens disappeared from the infected leaves within 3 days of spraying of this product.

Kim et al. (2009) have investigated the antifungal activity of three different types of silver nanoparticles against the fungus *Raffaelea* sp., which was respon-sible for the mortality of a large number of oak trees in Korea. Growth of fungi in the presence of silver nanoparticles was significantly inhibited. Effectiveness of combination of different forms of nanoparticles was also studied. It was found that silver nanoparticles caused detrimental effect not only on fungal hyphae but also on conidial germination. Copper nanoparticles in soda-lime glass powder showed efficient antimicrobial activity against gram-positive and gram-negative bacteria, as well as fungi (Esteban-Tejeda et al. 2009). According to Jo et al. (2009), silver nanoparticles were effective against plant pathogenic fungi such as *Bipolaris sorokiniana* and *M. grisea*. Similarly, Gajbhiye et al. (2009) reported the anti-fungal activity of silver nanoparticles against most important plant pathogenic fungi such as Fusarium, Phoma.

Nanopesticides, nanofungicides, and nanoherbicides are also being used in agriculture (Owolade and Ogunleti 2008). Many companies have made formula-tions which contain nanoparticles within the 100–250 nm size range that are able to dissolve in water more effectively than existing ones (thus increasing their activity). Some other companies have employed suspensions of nanoscale particles (nanoemulsions), which can be either water or oil based and contain uniform suspensions of pesticidal or herbicidal nanoparticles in the range of 200–400 nm. These have multiple applications in preventative measures and in the treatment or preservation of the harvested product (Goswami et al. 2010; Rickman et al. 2003).

References

Barik T, Sahu B, Swain V (2008) Nanosilica—from medicine to pest control. Parasitol Res 103(2):253–258

Bhattacharyya A, Bhaumik A, Rani PU, Mandal S, Epidi TT (2010) Nano-particles—a recent approach to insect pest control. Afr J Biotechnol 9(24):3489–3493

Cioffi N, Torsi L, Ditaranto N, Sabbatini L, Zambonin PG, Tantillo G, Ghibelli L, DAlessio M, Bleve-Zacheo T, Traversa E (2004) Antifungal activity of polymer-based copper nanocom-posite coatings. Appl Phys Lett 85(12):2417–2419

Esteban-Tejeda L, Malpartida F, Esteban-Cubillo A, Pecharromán C, Moya J (2009) Antibac-terial and antifungal activity of a soda-lime glass containing copper nanoparticles. Nanotechnology 20(50):505701

Gajbhiye M, Kesharwani J, Ingle A, Gade A, Rai M (2009) Fungus-mediated synthesis of silver nanoparticles and their activity against pathogenic fungi in combination with fluconazole. Nanomed Nanotechnol Biol Med 5(4):382–386

Goswami A, Roy I, Sengupta S, Debnath N (2010) Novel applications of solid and liquid formulations of nanoparticles against insect pests and pathogens. Thin Solid Films 519(3):1252–1257

Jo Y-K, Kim BH, Jung G (2009) Antifungal activity of silver ions and nanoparticles on phytopathogenic fungi. Plant Dis 93(10):1037–1043

Joseph T, Morrison M (2006) Nanotechnology in agriculture and food: a nanoforum report. Nanoforum.org

Kim SW, Kim KS, Lamsal K, Kim Y-J, Kim SB, Jung M, Sim S-J, Kim H-S, Chang S-J, Kim JK (2009) An in vitro study of the antifungal effect of silver nanoparticles on oak wilt pathogen Raffaelea sp. J Microbiol Biotechnol 19:760–764

Liu F, Wen L-X, Li Z-Z, Yu W, Sun H-Y, Chen J-F (2006) Porous hollow silica nanoparticles as controlled delivery system for water-soluble pesticide. Mater Res Bull 41(12):2268–2275

Owolade OF, Ogunleti DO (2008) Effects of titanium dioxide on the diseases, development and yield of edible cowpea. J Plant Prot Res 48(3):329–336

Park H-J, Kim S-H, Kim H-J, Choi S-H (2006) A new composition of nanosized silica—silver for control of various plant diseases. Plant Pathol J 22(3):295–302

Rai M, Ingle A (2012) Role of nanotechnology in agriculture with special reference to management of insect pests. Appl Microbiol Biotechnol 94(2):287–293

Rickman D, Luvall J, Shaw J, Mask P, Kissel D, Sullivan D (2003) Precision agriculture: changing the face of farming. Geotimes 48(11):28–33

Scrinis G, Lyons K (2007) The emerging nano-corporate paradigm: nanotechnology and the transformation of nature, food and agri-food systems. Int J Sociol Food Agri 15(2):22–44

Stadler T, Buteler M, Weaver DK (2010) Novel use of nanostructured alumina as an insecticide. Pest Manage Sci 66(6):577–579

Tungittiplakorn W, Cohen C, Lion LW (2004) Engineered polymeric nanoparticles for bioremediation of hydrophobic contaminants. Environ Sci Technol 39(5):1354–1358. doi:10.1021/es049031a

Ulrichs C, Mewis I, Goswami A (2005) Crop diversification aiming nutritional security in West Bengal: biotechnology of stinging capsules in nature's water-blooms. Ann Tech Issue State Agri Technol Serv Assoc ISSN 1–18

Wang L, Li X, Zhang G, Dong J, Eastoe J (2007) Oil-in-water nanoemulsions for pesticide formulations. J Colloid Interface Sci 314(1):230–235

Yang F-L, Li X-G, Zhu F, Lei C-L (2009) Structural characterization of nanoparticles loaded with garlic essential oil and their insecticidal activity against Tribolium castaneum (Herbst) (Coleoptera: Tenebrionidae). J Agri Food Chem 57(21):10156–10162

Chapter 9
A Brief Overview of Nanotechnology

Nanotechnology has emerged in the last decades of the twentieth century with the development of new enabling technologies for imaging, manipulating, and simulating matter at the atomic scale. The frontier of nanotechnology research and development encompasses a broad range of science and engineering activities directed toward understanding and creating improved materials, devices, and systems that can exploit the properties of matter emerging at the nanoscale. The results promise benefits that can shift paradigms in biomedicine (e.g., imaging, diagnosis, treatment, and prevention); energy (e.g., conversion and storage); electronics (e.g., computing and displays); manufacturing; environmental remediation; and many other categories of products and applications.

Among leading scientists, there is a growing awareness about the tremendous impact this field will have on the society and on the economy. It is forecasted to become possibly even more important than for example the invention of the steam engine or the discovery of penicillin.

The landmark lecture by eminent Nobel Laureate Richard Feynman in 1959 entitled "There's plenty of room at the bottom" brought life to the concept of nanotechnology, a technology which has been influencing different fields of research starting from hard-core science such as chemistry and physics to other applied fields of science such as electronics, materials science, biomedical science, agrochemicals, medicine and pharmaceutical science, etc. (Feynman 1960).

Nanotechnology and nanoscience are widely seen as having a great potential in bringing about benefits to many areas of research and applications. It is attracting increasing investments from governments and private sector businesses in many parts of the world. Concurrently, the application of nanoscience is raising new challenges in the safety, regulatory, and ethical domains, and this will require extensive debates on all levels.

The prefix nano is derived from a Greek word meaning dwarf. One nanometer (nm) is equal to one-billionth of a meter (that is 10^{-9} m). The term "nanotechnology" was first used in 1974, when Norio Taniguchi, a scientist at the University of Tokyo, Japan, referred to materials in nanometers.

At the nanometer scale, the physical, chemical, and biological properties of nanomaterials are fundamentally different from those of individual atoms,

A. De et al., *Targeted Delivery of Pesticides Using Biodegradable Polymeric Nanoparticles*, SpringerBriefs in Molecular Science, DOI: 10.1007/978-81-322-1689-6_9, © The Author(s) 2014

molecules, and bulk materials. They differ significantly from other materials due to two major factors: the increased surface area and quantum effects. A larger surface area usually results in more reactive chemical properties and also affects the mechanical or electrical properties of the materials. At the nanoscale, the quantum effects dominate the behaviors of a material, affecting its optical, electrical, and magnetic properties. By exploiting these novel properties, the main purpose of research and development in nanotechnology is to understand and create materials, devices, and systems with improved characteristics and performance (Thassu et al. 2007).

References

Feynman RP (1960) There's plenty of room at the bottom. Eng Sci 23(5):22–36
Thassu D, Deleers M, Pathak Y (eds) (2007) Nanoparticulate drug delivery systems, vol 166, 1st edn., Drugs and the pharmaceutical sciencesInforma Healthcare, New York

Chapter 10
Nanoparticulate Delivery Systems

A nanoparticulate system, typically, comprises particles or droplets in the submicron range, i.e., below 1 μm, in an aqueous suspension or emulsion, respectively. This small size of the inner phase gives such a system unique properties in terms of appearance and application. The particles are too small for sedimentation and they are held in suspension by the Brownian motion of the water molecules. They have a large overall surface area, and their dispersions provide a high solid content at low viscosity.

Historically, the first nanoparticles proposed as carriers for therapeutic applications were made of gelatin and cross-linked albumin (Scheffel et al. 1972; Marty et al. 1978). Since the use of proteins could stimulate the immune system, and moreover, to limit the toxicity of the cross-linking agents, nanoparticles made from synthetic polymers were developed. At first, the nanoparticles were made by emulsion polymerization of acrylamide and by dispersion polymerization of methyl methacrylate (Birrenbach and Speiser 1976; Kreuter and Speiser 1976). These nanoparticles were proposed as adjuvants for vaccines. Couvreur et al. (1979) proposed to make nanoparticles by polymerization of monomers from the family of alkylcyanoacrylates already used in vivo as surgical glue. During the same period of time, Gurny et al. (1981) proposed a method for synthesis of nanoparticles from another biodegradable polymer consisting of poly(lactic acid) used as surgical sutures in humans. Based on these initial investigations, several groups improved and modified the original processes mainly by reducing the amount of surfactant and organic solvents. A breakthrough in the development of nanoparticles occurred in 1986 with the development of methods allowing the preparation of nanocapsules corresponding to particles displaying a core–shell structure with a liquid core surrounded by a polymer shell (Al Khouri Fallouh et al. 1986; Legrand et al. 1999). Subsequently, the nanoprecipitation technique was proposed as well as the first method of interfacial polymerization in inverse microemulsion (Gasco and Trotta 1986). In the succeeding years, the methods based on salting-out (Allémann et al. 1992), emulsion–diffusion (Quintanar-Guerrero et al. 1998, 1999), and double emulsion techniques (Zambaux et al. 1998) were described. Finally, during the last decade, new approaches were considered to develop nanoparticles made from natural origin such as polysaccharides (Janes

A. De et al., *Targeted Delivery of Pesticides Using Biodegradable Polymeric Nanoparticles*, SpringerBriefs in Molecular Science, DOI: 10.1007/978-81-322-1689-6_10, © The Author(s) 2014

et al. 2001; Prabaharan and Mano 2005; Liu et al. 2008). These nanoparticles were mainly developed for peptides and nucleic acid delivery. Further development such as surface modification of nanoparticles was carried out later (Gref et al. 1997).

10.1 Delivery Systems

The specific delivery of active principles to the target site, organ, a tissue, or unhealthy cells by carriers is one of the major challenges in the delivery of bioactive molecules. Many of the bioactive compounds have physicochemical characteristics that are not favorable to transit through the biological barriers that separate the administration site from the site of action. Some of the active compounds run up against enzymatic barriers, which can lead to their degradation and fast metabolization. Therapeutically, distribution of such active molecules to the diseased target zones can therefore be difficult. Moreover, the accumulation of molecules in healthy tissues can cause unacceptable toxic effects, leading to the abandonment of treatment despite their effectiveness (Soussan et al. 2009).

In order to overcome the above challenges, an ideal delivery system must possess basically two elements—the ability to transport loaded payload to the target site and control their release. The targeting can ensure high efficiency of delivery of the payload at the site of core interest and reduce any unwanted biological effects. Various delivery devices have been developed, and an overview of each type of nanocarriers is given in the following section.

According to the process used for the preparation of nanoparticles, nanospheres or nanocapsules have been obtained. Nanospheres are homogeneous matrix systems in which the active ingredient is dispersed throughout the particles. Nanocapsules are vesicular systems in which the molecules are confined to a cavity surrounded by a polymeric membrane (Lamprecht 2009).

10.1.1 Hydrogel

Hydrogels are three-dimensional networks composed of hydrophilic polymer chains. They have the ability to swell in water without dissolving. The type of cross-linking between the polymer chains can be chemical (covalent bonds) or physical (hydrogen bonds or hydrophobic interactions). The high water content in these materials makes them highly biocompatible. There are natural hydrogels such as DNA, proteins, or synthetic [e.g., poly(2-hydroxyethyl methacrylate), poly(N-isopropylacrylamide)] or a biohybrid (Peppas et al. 2006; Letchford and Burt 2007). The release mechanism can be induced by temperature or pH.

Temperature-controlled release can be due to the competition between hydrogen bonding and hydrophobic interactions. At lower temperatures, the hydrogen bonding between polar groups of the polymer can be predominant causing the polymer to swell in water. At higher temperatures, the hydrophobic interactions can takeover, leading to shrinkage (Bae et al. 1991a, b). In a similar manner, glucose-sensitive hydrogels can be made to release insulin in a controlled fashion in response to the demand (Yuk et al. 1997).

10.1.2 Dendrimers

Dendrimers are highly branched cascade molecules that emanate from a central core through a stepwise repetitive reaction sequence. Such molecules consist of three topologically different regions: a small initiator core of low density and multiple branching units (the density of which increases with increasing distance from core), eventually leading to a rather densely packed shell. The outer terminal units for shielding can actually amount to an encapsulation that can create a distinct microenvironment around the core moieties and hence affect their properties (Hecht and Fréchet 2001).

Dendrimers can be synthesized in multiple ways. A dendrimer can be synthesized originating form core by repetition of a sequence of reactions, which allows fast growth of the dendrimer in both size and number of terminal groups (Tomalia et al. 1985). In another convergent method, the core is incorporated in the final step of elaboration of the dendrimer (Hawker and Frechet 1990).

Owing to their large number of surface groups, dendrimers can have the ability to create multivalent interactions (Mammen et al. 1998). Dendritic structures may also be engineered to encapsulate certain hydrophobic molecules (like indomethacin) as well (Liu et al. 2000).

The dendrimeric surface can be tuned for functional groups so as to induce an electrostatic-type interaction with the active molecules. As an example, negatively charged DNA chains can be complexed to positively charged dendrimers. Several research groups have demonstrated that dendrimer/DNA complexes (which are very compact) can easily penetrate cells by endocytosis and, therefore, improve the transfection efficiency (Tang et al. 1996; Zinselmeyer et al. 2002; Loup et al. 1999; Caminade et al. 2008; Tang and Szoka 1997). In some cases, the bulkiness of the dendrimer and the density of their structure can make cleavage of water-soluble and biodegradable bonds of the peripheral layer quite difficult (Jansen et al. 1995). Delivery of active principles is therefore not so straightforward in these cases. In other cases, the encapsulated molecules may not be well trapped and may be released prematurely (Liu et al. 2000). Nevertheless, the functional groups of dendrimers can be easily tuned, making them versatile molecular carriers.

10.1.3 Liposomes

Liposomes are vesicles formed by the auto-association of one or several phospholipid bilayers that can enclose an aqueous compartment. They have attracted the attention of a number of research groups in various fields, such as physical chemistry, biophysics, and pharmaceutics because of their structure (which is comparable to the phospholipid membranes of living cells) (Samad et al. 2007). The innocuous nature of phospholipidic components in liposomes makes them suitable reservoir systems that have rapidly become ideal candidates for molecular vectorization in biological media. In general, liposomes are able to transport both hydrophobic substances anchored into their bilayer and hydrophilic substances encapsulated in their cavity. Temperature-sensitive liposomes have also been elaborated using lipids such as 1,2-dipalmitoyl-sn-glycero-3-phosphocholine, which has a phase-transition temperature between 41 and 43 °C. These liposomes can be used in association with hyperthermia treatments (for example in the delivery of drugs into solid tumors) (Needham and Dewhirst 2001). Liposomes have been used extensively in pharmaceutical research. Ligands can be anchored onto the surface of liposomes, so as to deliver encapsulated drugs to specific action sites. These ligands can be antibodies, which can bind to specific cell receptors, or less-specific ligands, such as folate or selectin (Forssen and Willis 1998). Attachment of PEG to liposomes can also protect them from detection by monocytes and macrophages (Gabizon 1992) in the liver and spleen (allowing a prolonged circulation time within the bloodstream). The liposomes utilized in doxil (which is marketed as a chemotherapy drug) have been formulated with surface-bound methoxypolyethylene glycol (MPEG). Liposomes are thus versatile reservoir systems. The more they develop the more sophisticated their compositions become, allowing very specific targeting and completely controlled drug delivery. However, these are rather complex systems which have to be systematically tuned according to the drug to be encapsulated and also according to the desired application.

The physical and chemical instability of liposomes also limits their use in vectorization. Chemically, their poor stability can be attributed to their ester bond hydrolysis, and physically, the aggregation or the fusion of several liposomes can lead to the formation of large-sized entities that are no longer usable in vectorization. Moreover, the liposomes may be subject to leakage, releasing the encapsulated drugs before they can reach their site of action. Their preparation procedure also requires the use of an organic solvent, which can always leave toxic residual traces.

10.1.4 Niosomes

Niosomes (Soussan et al. 2009) are made of nonionic surfactants that are organized into spherical bilayers enclosing an aqueous compartment and have an identical structure to liposomes and polymersomes. Several preparation methods for niosomes have been described in the literature (Uchegbu and Vyas 1998). In most cases, niosome formation requires the addition of molecules such as cholesterol to stabilize the bilayer and molecules that prevent the formation of niosome aggregates by steric or electrostatic repulsion.

In an analogous fashion to liposomes, niosomes are able to vectorize hydrophobic drugs enclosed in their bilayer and hydrophilic substances encapsulated in their aqueous cavity. Unlike phospholipidic liposomes, niosomes, which are made of surfactants, are not sensitive to hydrolysis or oxidation. This is an advantage for their use in biological media. Moreover, surfactants are cheaper and easier to store than phospholipids. A further advantage of niosomes relative to liposomes lies in their formulation, as these vectors can be elaborated from a wide variety of surfactants, the hydrophilic heads of which can be chosen according to the application and the desired site of action (Uchegbu and Vyas 1998). Notably, surfactant niosomes have been obtained with glycerol (Lesieur et al. 1990), ethylene oxide (Gianasi et al. 1997), crown ethers (Darwish and Uchegbu 1997), and polyhydroxylated (Assadullahi et al. 1991) or sugar-based (Polidori et al. 1994) polar headgroups.

The encapsulation of active substances in niosomes has been found to reduce their toxicity, increase their absorption through cell membranes, and allow them to target organs or specific tissues inside the body. Recently, antibody surface-functionalized niosomes have been developed in a similar manner as virosomes (Hood et al. 2007).

Niosomes have been developed so as to achieve the same specific drug delivery objectives as liposomes. However, niosome membranes are permeable to low-molecular-weight molecules and a leakage of drugs encapsulated in the aqueous cavity of niosomes over time cannot be ruled out.

10.1.5 Polymersomes

Polymersomes are tanklike systems consisting of a liquid central core enclosed in a thin polymer wall not more than a few nanometers thick (Letchford and Burt 2007; Rijcken et al. 2007; Meng et al. 2005). The polymersome membrane can be formed from a block copolymer that is organized in a bilayer, in a similar fashion to those of the liposomes. These polymersomes thus have an aqueous internal cavity. Polymersomes can exhibit versatile transport properties, as hydrophobic drugs can be enclosed in the membrane of the carrier, whereas hydrophilic drugs can be encapsulated in their aqueous cavity.

Polymersome systems have been used for the delivery of anticancer drugs, such as paclitaxel (hydrophobic) and doxorubicin (hydrophilic). Doxorubicin has been encapsulated in the internal cavity of the polymersome, whereas paclitaxel has been incorporated into the polymer bilayer during the polymer film formation so as to maximize the anticancer drug efficiency (Ahmed et al. 2006). Polymersomes have been were obtained by mixing two block copolymers, namely biodegradable

PLA-PEG and inert poly(ethylene glycol)–poly(butadiene) (PEG-PBD). Hydrolysis of PLA-PEG can form pores in the membrane, which can allow the delivery of both the drugs to be controlled. It has been found that the combination therapy with doxorubicin–paclitaxel-loaded polymersomes can trigger massive apoptosis in tumor mass within first day of treatment and there is twice as much enhancement of apoptosis as compared to free drug therapy.

Despite their efficiency, the major drawback of polymersomes is their instability, leading to leakage of the encapsulated drugs. Moreover, passive encapsulation (used in the case of polymersomes) requires a high amount of active substances (as the encapsulated concentration is identical to the concentration of the aqueous solution used to rehydrate the polymer film).

10.1.6 Solid Lipid Nanoparticle

Nanoparticles (Muller et al. 2000) composed of lipids, which are solid at room and physiologic temperatures, are referred to as solid lipid nanoparticle (SLN). These are typically composed of stabilizing surfactants, triglycerides, glyceride mixtures, and waxes. They are usually prepared by various procedures like high-pressure homogenization, microemulsion, and nanoprecipitation. Generally, lipids such as triglycerides are well tolerated by the organism. Moreover, the production of these nanoparticles is much simpler than that of the nanospheres and can be transposed to the industrial scale at a lower cost.

Typically, the active substance required for the desired application is dissolved or dispersed into the molten lipid phase. Following fast cooling of the glycerides, an α-crystalline structure is obtained that is unstable and not well ordered (Bunjes et al. 1996). Active molecules then preferentially gather in the amorphous areas of the matrix. However, the α-crystalline structure adopted by the lipids alters during standing to a β-crystalline structure, which is more stable and better ordered (Westesen et al. 1993). During this rearrangement, the increase in the ordering of the lipid phase leads to an expulsion of the active substances into the amorphous regions (Pietkiewicz et al. 2006). Control of the lipid matrix transformation from the α-form to the β-form (for example, by temperature control) should therefore allow an on-command release of the drug (Muller et al. 2000). However, to date, these SLN with controlled crystalline transformation have not been fully mastered.

As the drug-loading capacity of the particles relies essentially on the structure and the polymorphism of the lipid forming the nanoparticles, some new types of lipid particles exhibiting amorphous zones have been developed (Müller et al. 2002a; Wissing et al. 2004). These lipid particles, which are partially crystalline, can be composed of a mixture of glycerides with different fatty acids possessing various chain lengths and degree of unsaturation, leading to an imperfect material (and therefore offering a better drug-loading rate). A second type of lipid particle, called multiple lipid particle, is obtained by mixing liquid lipids with solid lipids when preparing the nanoparticles. The active substances can become localized in the oily compartments contained in the solid lipid particles. Finally, an amorphous system can be obtained with a particular mixture of lipids. The incorporation of active molecules into this kind of solid nanoparticles is one of the most efficient.

The use of these solid nanoparticles in drug vectorization is now under development, as both in vitro and in vivo studies have proved that these carriers are well tolerated. However, the polymorphism of these lipid matrixes and possible crystal rearrangements has to be controlled to avoid stability problems in these structures (gelification problems) (Mehnert and Mäder 2001). Moreover, the release of the active molecules incorporated in these solid nanoparticles is not always well controlled, and this can limit their applications.

10.1.7 Micro- and Nanoemulsions

Emulsions are heterogenous dispersions of two immiscible liquids such as oil-in-water (O/W) or water-in-oil (W/O). They typically require a surfactant, as without surfactant molecules, they are susceptible to rapid degradation by coalescence or flocculation, leading to phase separation (Fukushima et al. 2000). The use of micro- and nanoemulsions are becoming increasingly common in drug delivery systems. Microemulsion is used to denote a thermodynamically stable, fluid, transparent (or translucent) dispersion of oil and water, stabilized by an interfacial film of amphiphilic molecules (Danielsson and Lindman 1981). The striking difference between a conventional emulsion (1–10 μm) and the microemulsion (200 nm–1 μm) is that the latter does not need any mechanical input for their formation as they are thermodynamically more stable. On the other hand, nanoemulsions (20–200 nm) are kinetically more stable.

Nanoemulsions are of great interest in pharmaceutical, cosmetic formulations (Solans et al. 2005). Nanoemulsions are used as drug delivery systems for administration through various systemic routes. Parenteral administration (Tamilvanan et al. 2005) of nanoemulsions is employed for a variety of purposes, including controlled drug delivery of vaccines or as gene carriers (Tamilvanan 2004; Pan et al. 2003). The benefit of nanoemulsions in the oral (Nicolaos et al. 2003) and ocular (Tamilvanan 2004; Rabinovich-Guilatt et al. 2004) administration of drugs has been also reported. Cationic nanoemulsions have been evaluated as DNA vaccine carriers (Bivas-Benita et al. 2004). They are also interesting candidates for the delivery of drugs or DNA plasmids through the skin after topical administration (Fang et al. 2004; Wu et al. 2001). The drawback in an emulsion system can be the use of high concentration of surfactant, leading to toxicity.

10.1.8 Micelles

Micelles are aggregates of amphiphilic molecules in which the polar headgroups are in contact with water, and the hydrophobic moieties are gathered in the core so as to minimize their contact with water. The main driving force in the auto-association process of these surfactant molecules is their hydrophobicity. The micelles form above a certain concentration, known as the critical micelle concentration (CMC). The mean size of these objects usually varies from 1 to 100 nm. The micellar systems are dynamic in nature, as the surfactants can collide and exchange their contents freely and rapidly.

In addition to surfactants, block copolymers (having both a hydrophilic and a hydrophobic part) or triblock copolymers (with one hydrophobic and two hydrophilic parts or one hydrophilic and two hydrophobic parts) can also self-assemble to form polymeric micelles. These polymeric micelles can have a mean diameter of 20–50 nm and are practically monodisperse. Polymeric micelles are generally more stable than surfactant micelles and they form at markedly lower CMCs. These objects are also much less dynamic than those formed from surfactants.

Polymeric micelles are more frequently used in vectorization than surfactant micelles. The slow degradation kinetics of polymeric micelles has contributed to their success in vectorization applications, usually for anticancer hydrophobic drug (such as paclitaxel) delivery to tumors.

Polymeric micelles also have the advantage of being able to deliver an active principle to a specific site of action, provided the polymer structure is tuned properly. An example is the development of pH-sensitive copolymers by inclusion of amine (Martin et al. 1996) or acid functional groups (Mitsukami et al. 2001) into the copolymer skeleton. The active principles can then be delivered by micelle destabilization at a site of action possessing a specific pH.

The major drawback of micellar vectors (and in particular surfactant vectors) is their tendency to break up upon dilution. This is not the case for polymeric micelles, but in this case their synthesis can sometimes prove difficult.

10.1.9 Carbon Nanomaterials

Carbon nanomaterials for drug delivery applications mainly include fullerenes and carbon nanotubes (single and multiwalled). Considerable amount of work has been done to utilize them as nanocarriers for drug delivery (Bianco and Prato 2003; Bianco et al. 2005; Kam and Dai 2005). The inert surface of these materials has posed challenges in terms of surface modifications and in making them water soluble, biocompatible, and fluorescent. But despite all these, a number of recent reports establish that carbon nanotubes can be toxic (Jia et al. 2005). More recently, glucose-derived functionalized carbon spheres have shown hopes to behave as efficient nanocarriers (Selvi et al. 2008). More detailed studies on their mechanism of entry and other possible applications are still awaited.

10.2 Synthesis of Nanomaterials

The approaches for synthesis of nanomaterials are commonly categorized into top-down approach, bottom-up approach, and hybrid approach.

10.2.1 Top-Down Approach

This approach starts with a block of material and reduces the starting material down to the desired shape in nanoscale by controlled etching, elimination, and layering of the material. An example includes a nanowire fabricated by lithography impurities and structural defects on the surface. One problem with the top-down approach is the imperfections of the surface structure, and this may significantly affect the physical properties and surface chemistry of the nanomaterials. Further, some uncontrollable defects may also be introduced even during the etching steps. Regardless of the surface imperfections and other defects, the top-down approach is still one of the important ways for synthesizing nanomaterials (Teli et al. 2010). This technique employs two very common high-energy shear force methods, viz. milling and high-pressure homogenization (Swarbrick 2006). Milling yields nanoparticle in dry state, and high-pressure homogenization produces them in a suspension form.

10.2.2 Bottom-Up Approach

In a bottom-up approach, materials are fabricated by efficiently and effectively controlling the arrangement of atoms, molecules, macromolecules, or supramolecules. The synthesis of large polymer molecules is a typical example of the bottom-up approach, where individual building blocks (monomers) are assembled into a large molecule (or polymerized into bulk material). The main challenge for the bottom-up approach is how to fabricate structures which are of sufficient size and amount (to be used as materials in practical applications). Nevertheless, the nanostructures fabricated in the bottom-up approach usually have fewer defects, a more homogeneous chemical composition and better short and long range ordering (Teli et al. 2010). In the bottom-up approach precipitation, crystallization and single droplet evaporation processes are typically used produce nanoparticles (Chan and Kwok 2011). Some of the techniques used for fabrication of nanoparticles using the bottom-up approach are detailed later in further sections.

10.2.3 Hybrid Approach

Though both the top-down and bottom-up approaches play important roles in the synthesis of nanomaterials, some technical problems exist with these two approaches. It is found that, in many cases, combining top-down and bottom-up method into an unified approach can transcend the limitations of both so as to give an optimal solution (Teli et al. 2010). A thin film device, such as a magnetic sensor, is usually developed in a hybrid approach, since the thin film is grown in a bottom-up approach, whereas it is etched into the sensing circuit in a top-down approach.

10.3 Dispersion of Preformed Polymers

10.3.1 Emulsification/Solvent Evaporation

A hydrophobic polymer can be dispersed in an organic solution to form nano-droplets, using a dispersing agent and high-energy homogenization (Tice and Gilley 1985), in a non-solvent or suspension medium such as chloroform, dichloromethane (ICH, class 2), or ethyl acetate (ICH, class 3) (ICH-Guideline 2011). The polymer can then precipitate in the form of nanospheres, and the molecule to be entrapped can be finely dispersed in the polymer matrix network. The solvent can be subsequently evaporated by increasing the temperature under pressure or by continuous stirring and the size of the nanospheres can be controlled by adjusting the stir rate, type and amount of the dispersing agent, viscosity of the organic and aqueous phases, and temperature (Pinto Reis et al. 2006). In the conventional methods, two main strategies are typically used for the formation of emulsions: the preparation of single emulsions, e.g., oil-in-water (o/w) or double-emulsions, e.g., (water-in-oil)-in-water (w/o)/w (Rao and Geckeler 2011). Even though different types of emulsions may be used, oil/water emulsions are of interest because they use water as the non-solvent; this simplifies and thus improves process economics, because it eliminates the need for recycling, facilitates the washing step, and minimizes agglomeration. However, this method can only be applied to liposoluble molecules, and limitations are imposed by the scale-up of the high-energy requirements in homogenization. Frequently used polymers are PLA (Ueda and Kreuter 1997), PLGA (Tabata and Ikada 1989), PCL (Gref et al. 1994), and poly(h-hydroxybutyrate) (Koosha et al. 1989). Some of the molecules encapsulated by this method are tetanus toxoid (Tobío et al. 1998), loperamide (Ueda and Kreuter 1997), and cyclosporin A (Jaiswal et al. 2004).

10.3.2 Solvent Displacement, Diffusion, or Nanoprecipitation

A solution of polymer, molecule to be entrapped (a drug) and lipophilic stabilizer (surfactant) in a semi-polar solvent (i.e., miscible with water), is injected into an aqueous solution (that is a non-solvent or anti-solvent for the molecule to be entrapped and the polymer) containing another stabilizer under moderate stirring. Nanoparticles can be formed instantaneously by rapid solvent diffusion. The organic solvent can be removed subsequently under reduced pressure. The velocity of solvent removal (and thus nuclei formation) can be the key to obtain particles in the nanometer range instead of larger lumps or agglomerates (Lamprecht 2009). As an alternative to liquid organic or aqueous solvents, supercritical fluids can also be applied. Fessi et al. (1989) proposed a simple and mild method yielding nanoscale and monodisperse polymeric particles without the use of any pre-liminary emulsification for encapsulation of indomethacin. In this, both the sol-vent, solvent and the non-solvent must have low viscosity and high mixing capacity in all proportions, such as acetone (ICH, class3) (ICH-Guideline 2011) and water. In addition, another delicate parameter can be the composition of the solvent/polymer/water mixture limiting the feasibility of the nanoparticle forma-tion. The only complementary operation following the mixing of the two phases is the removal of the volatile solvent by evaporation under reduced pressure. One of the most interesting and practical aspects of this method is its capacity to be scaled up from laboratory to industrial amounts, since the process can be run with con-ventional equipment.

This method has been applied to various polymeric materials such as PLA (Némati et al. 1996) and PCL (Molpeceres et al. 1996). Barichello et al. (1999) have shown the application of this method for the entrapment of valproic acid, ketoprofen, vancomycin, phenobarbital, and insulin by using PLGA polymer.

10.3.3 Emulsification/Solvent Diffusion

In this method, the encapsulating polymer is dissolved in a partially water-soluble solvent such as propylene carbonate and saturated with water to ensure the initial thermodynamic equilibrium of both the liquids. In order to produce the precipi-tation, it is necessary to promote the diffusion of the solvent of the dispersed phase by dilution with an excess of water when the organic solvent is partly miscible with water (or with another organic solvent in the opposite case). Subsequently, the polymer–water-saturated solvent phase is emulsified in an aqueous solution containing stabilizer, leading to solvent diffusion to the external phase and the formation of nanospheres or nanocapsules, according to the oil-to-polymer ratio. Finally, the solvent is eliminated (Pinto Reis et al. 2006). Several drug-loaded nanoparticles have been produced by the emulsification/solvent diffusion (ESD) technique, including doxorubicin–PLGA conjugate nanoparticles (Yoo et al. 1999), plasmid DNA-loaded PLA-PEG nanoparticles (Perez et al. 2001), cyclosporin A (Cy-A)-loaded gelatin and cyclosporin (Cy-A)-loaded sodium glycolate nanoparticles (El-Shabouri 2002).

10.3.4 Salting-Out

Salting-out is based on the separation of a water-miscible solvent from an aqueous solution via a salting-out effect. The salting-out procedure can be considered as a modification of the emulsification/solvent diffusion technique (Pinto Reis et al. 2006; Rao and Geckeler 2011). The polymer and the molecule to be entrapped are initially dissolved in a solvent such as acetone (ICH, class 3), which is subsequently emulsified into an aqueous gel containing the salting-out agent (electrolytes, such as magnesium chloride, calcium chloride, and magnesium acetate, or non-electrolytes, such as sucrose) and a colloidal stabilizer such as polyvinylpyrrolidone or hydroxyethylcellulose. This oil/water emulsion is diluted with a sufficient volume of water or aqueous solution to enhance the diffusion of acetone into the aqueous phase, thus inducing the formation of nanospheres. The selection of the salting-out agent is important, because it can play an important role in the encapsulation efficiency. Both the solvent and the salting-out agent can then be eliminated by cross-flow filtration.

In a work carried out by Song et al., PLGA nanoparticles have been prepared by employing NaCl as the salting-out agent instead of $MgCl_2$ or $CaCl_2$ (Song et al. 2008).

10.3.5 Dialysis

In this method, the polymer is dissolved in an organic solvent and placed inside a dialysis bag with a proper molecular weight cutoff. Dialysis is performed against a non-solvent miscible with the organic solvent. The displacement of the solvent inside the membrane is followed by the progressive aggregation of polymer due to a loss of solubility, leading to the formation of a homogeneous suspension of nanoparticles. The dialysis method has been used for synthesizing PLGA (Choi and Kim 2007), PLA (Liu et al. 2007), and dextran ester (Hornig and Heinze 2007) nanoparticles. Poly(ε-caprolactone)-grafted poly(vinyl alcohol) copolymer nanoparticles (Sheikh et al. 2009) have been investigated using this method as drug carrier models for hydrophobic and hydrophilic anti-cancer drugs paclitaxel and doxorubicin. In vitro drug release experiments have been conducted, and the loaded nanoparticles revealed continuous and sustained release form for both the drugs, up to 20 and 15 days for paclitaxel and doxorubicin, respectively.

10.3.6 Supercritical Fluid Technology

Conventional methods such as in situ polymerization and solvent evaporation often require the use of toxic solvents and surfactants. Supercritical fluids allow attractive alternatives for the nanoencapsulation process because these are environment-friendly solvents. The commonly used methods of supercritical fluid technology are the rapid expansion of supercritical solution (RESS) and the supercritical anti-solvent (SAS) methods (Cristian and Karol 2009; Rao and Geckeler 2011). A supercritical fluid is a substance that is used in a state above the critical temperature and pressure where gases and liquids can coexist. It is able to penetrate materials like a gas and is also able to dissolve materials like a liquid. As an example, use of carbon dioxide or water in the form of a supercritical fluid can allow substitution for an organic solvent.

In the RESS method, a polymer is solubilized in a supercritical fluid and the solution is expanded through a nozzle. The solvent power of the supercritical fluid dramatically decreases and the solute eventually precipitates out. A uniform distribution of drug inside the polymer matrix (e.g., PLA nanospheres) can be achieved only for low molecular mass (<10,000) polymers because of the limited solubility of high molecular mass polymers in supercritical fluids. Chernyak et al. (2001) have produced droplets of poly(perfluoropolyether diamide) from the rapid expansion of CO_2 solutions. Sane and Thies have also presented this method for developing poly(l-lactide) nanoparticle by using CO_2 + THF solution (Sane and Thies 2007).

In the SAS method, the solution is charged with the supercritical fluid in the precipitation vessel containing a polymer in an organic solvent. At high pressure, enough anti-solvent can enter into the liquid phase so that the solvent power is lowered and the polymer precipitates. Following precipitation, the anti-solvent can flow through the vessel to strip the residual solvent. When the solvent content has been reduced to the desired level, the vessel can be depressured and the solid nanoparticles can be collected. Meziani et al. (2004) have reported the preparation of poly(heptadecafluorodecylacrylate) nanoparticles by this technique.

10.4 Emulsion Polymerization

Emulsion polymerization is the most common method used for the production of a wide range of specialty polymers. The use of water as a dispersion medium is not only environment-friendly, but it also allows excellent heat dissipation during the course of the polymerization. Based on the utilization of surfactant, the method can be classified as conventional and surfactant-free emulsion polymerization (Rao and Geckeler 2011).

10.4.1 Conventional Emulsion Polymerization

In conventional emulsion polymerization (Rao and Geckeler 2011), initiation occurs when a monomer molecule dissolved in the continuous phase, collides with an initiator molecule (that may be an ion or a free radical). Alternatively, the monomer molecule can be transformed into an initiating radical by high-energy radiation (including γ-radiation, ultraviolet, or strong visible light). Phase separation and formation of solid particles can take place before or after the termination of the polymerization reaction. Brush-type amphiphilic block copolymers of polystyrene-b-poly-(poly(ethylene glycol) methyl ether methacrylate) have been synthesized by the conventional emulsion polymerization method (Muñoz-Bonilla et al. 2010).

10.4.2 Surfactant-Free Emulsion Polymerization

This technique has received considerable attention for use as a simple and green process for nanoparticle production without the addition and subsequent removal of the stabilizing surfactants (Rao and Geckeler 2011). The reagents used in an emulsifier-free system include deionized water, a water-soluble initiator (potassium persulfate) and monomers (more commonly vinyl or acryl monomers). In such polymerization systems, stabilization of the nanoparticles can occur through the use of ionizable initiators or ionic co-monomers. The emulsifier-free monodisperse poly(methyl methacrylate) (PMMA) microspheres have been synthesized by this method using microwave irradiation (Bao and Zhang 2004). The emulsifier-free core–shell polyacrylate latex nanoparticles containing fluorine and silicon in shell have also been successfully synthesized by emulsifier-free seeded emulsion polymerization technique with water as the reaction medium (Cui et al. 2007).

10.4.3 Miniemulsion Polymerization

Miniemulsion polymerization technique uses water, monomer mixture, co-stabilizer, surfactant, and an initiator (Rao and Geckeler 2011). The key difference between emulsion polymerization and miniemulsion polymerization is the utilization of a low molecular mass compound as a co-stabilizer and also the use of a high-shear device (ultrasound, etc.). Miniemulsions are critically stabilized, require a high-shear device to reach a steady state and have an interfacial tension much greater than zero. Polymethylmethacrylate (Mouran et al. 1996) and poly(n-butylacrylate)

(Leiza et al. 1997) nanoparticles have been produced by employing sodium lauryl sulfate/dodecyl mercaptan and sodium lauryl sulfate/hexadecane as surfactant/ co-stabilizer systems, respectively.

10.4.4 Microemulsion Polymerization

In a microemulsion polymerization, an initiator (typically water soluble) is added to the aqueous phase of a thermodynamically stable microemulsion containing swollen micelles. The polymerization can start from this thermodynamically stable, spontaneously formed state and relies on high quantities of surfactant systems (which possess an interfacial tension at the oil/water interface close to zero). Furthermore, the particles are completely covered with surfactant because of the utilization of a high amount of surfactant (Rao and Geckeler 2011). Initially, polymer chains are formed only in some droplets, as the initiation cannot be attained simultaneously in all microdroplets. Later, the osmotic and elastic influence of the chains destabilizes the fragile microemulsions and typically leads to an increase in the particle size, the formation of empty micelles and secondary nucleation. Synthesis of a functional copolymer of methyl methacrylate and N-methylolacrylamide (NMA) (Macías et al. 1995) and polymerization of vinyl acetate (Sosa et al. 2000) in microemulsions have been prepared with the surfactant dioctyl sodium sulfosuccinate (aerosol OT). Aerosol OT has also been used to prepare the polymeric nanoparticles of polyvinylpyrrolidone cross-linked with N,N'-methylene bis-acrylamide in the aqueous cores of reverse micellar droplets as a nanoreactors. This technique has been used to prepare the hydrogel nanoparticles of diameter below 100 nm encapsulating water-soluble materials. The particles could be lyophilized and redispersed in aqueous buffer without changing their size and surface morphology (Bharali et al. 2003).

10.4.5 Interfacial Polymerization

Interfacial polymerization involves step polymerization of two reactive monomers or agents, which are dissolved, respectively, in two phases (i.e., continuous and dispersed phases), and the reaction can take place at the interface of the two liquids (Karode et al. 1998). The relative ease of obtaining interfacial polymerization has made it a preferred technique in many fields, ranging from the encapsulation of pharmaceutical products to the preparation of conducting polymers (Rao and Geckeler 2011). α-tocopherol-loaded polyurethane and poly(ether urethane)-based nanocapsules have been reported to be prepared by this method by Bouchemal et al. (2004). Core–shell biocompatible polyurethane nanocapsules encapsulating ibuprofen have been also been obtained by interfacial polymerization (Gaudin and Sintes-Zydowicz 2008).

References

Ahmed F, Pakunlu RI, Brannan A, Bates F, Minko T, Discher DE (2006) Biodegradable polymersomes loaded with both paclitaxel and doxorubicin permeate and shrink tumors, inducing apoptosis in proportion to accumulated drug. J Control Release 116(2):150–158

Al Khouri Fallouh N, Roblot-Treupel L, Fessi H, Devissaguet JP, Puisieux F (1986) Development of a new process for the manufacture of polyisobutylcyanoacrylate nanocapsules. Int J Pharm 28(2–3):125–132

Allémann E, Gurny R, Doelker E (1992) Preparation of aqueous polymeric nanodispersions by a reversible salting-out process: influence of process parameters on particle size. Int J Pharm 87(1–3):247–253

Assadullahi TP, Hider RC (1083) McAuley AJ (1991) Liposome formation from synthetic polyhydroxyl lipids. Biochim Biophys Acta (BBA): Lipids Lipid Metab 3:271–276

Bae YH, Okano T, Kim SW (1991a) "On–off" thermocontrol of solute transport. I. Temperature dependence of swelling of N-isopropylacrylamide networks modified with hydrophobic components in water. Pharm Res 8(4):531–537

Bae YH, Okano T, Kim SW (1991b) "On–off" thermocontrol of solute transport. II. Solute release from thermosensitive hydrogels. Pharm Res 8(5):624–628

Bao J, Zhang A (2004) Poly(methyl methacrylate) nanoparticles prepared through microwave emulsion polymerization. J Appl Polym Sci 93(6):2815–2820. doi:10.1002/app.20758

Barichello JM, Morishita M, Takayama K, Nagai T (1999) Encapsulation of hydrophilic and lipophilic drugs in PLGA nanoparticles by the nanoprecipitation method. Drug Dev Ind Pharm 25(4):471–476. doi:10.1081/DDC-100102197

Bharali DJ, Sahoo SK, Mozumdar S, Maitra A (2003) Cross-linked polyvinylpyrrolidone nanoparticles: a potential carrier for hydrophilic drugs. J Colloid Interface Sci 258(2):415–423

Bianco A, Prato M (2003) Can carbon nanotubes be considered useful tools for biological applications? Adv Mater 15(20):1765–1768. doi:10.1002/adma.200301646

Bianco A, Kostarelos K, Prato M (2005) Applications of carbon nanotubes in drug delivery. Curr Opin Chem Biol 9(6):674–679. doi:10.1016/j.cbpa.2005.10.005

Birrenbach G, Speiser PP (1976) Polymerized micelles and their use as adjuvants in immunology. J Pharm Sci 65(12):1763–1766

Bivas-Benita M, Oudshoorn M, Romeijn S, van Meijgaarden K, Koerten H, van der Meulen H, Lambert G, Ottenhoff T, Benita S, Junginger H, Borchard G (2004) Cationic submicron emulsions for pulmonary DNA immunization. J Control Release 100(1):145–155. doi:10.1016/j.jconrel.2004.08.008

Bouchemal K, Briançon S, Perrier E, Fessi H, Bonnet I, Zydowicz N (2004) Synthesis and characterization of polyurethane and poly(ether urethane) nanocapsules using a new technique of interfacial polycondensation combined to spontaneous emulsification. Int J Pharm 269(1):89–100. doi:10.1016/j.ijpharm.2003.09.025

Bunjes H, Westesen K, Koch MHJ (1996) Crystallization tendency and polymorphic transitions in triglyceride nanoparticles. Int J Pharm 129(1–2):159–173. doi:10.1016/0378-5173(95)04286-5

Caminade A-M, Turrin C-O, Majoral J-P (2008) Dendrimers and DNA: combinations of two special topologies for nanomaterials and biology. Chem: A Eur J 14(25):7422–7432. doi:10.1002/chem.200800584

Chan H-K, Kwok PCL (2011) Production methods for nanodrug particles using the bottom-up approach. Adv Drug Deliv Rev 63(6):406–416. doi:10.1016/j.addr.2011.03.011

Chernyak Y, Henon F, Harris RB, Gould RD, Franklin RK, Edwards JR, DeSimone JM, Carbonell RG (2001) Formation of perfluoropolyether coatings by the rapid expansion of supercritical solutions (RESS) process. Part 1: experimental results. Ind Eng Chem Res 40(26):6118–6126. doi:10.1021/ie010267m

Choi S-W, Kim J-H (2007) Design of surface-modified poly(d, 1-lactide-co-glycolide) nanoparticles for targeted drug delivery to bone. J Control Release 122(1):24–30. doi:10.1016/j.jconrel.2007.06.003

Couvreur P, Kante B, Roland M, Guiot P, Bauduin P, Speiser P (1979) Polycyanoacrylate nanocapsules as potential lysosomotropic carriers: preparation, morphological and sorptive properties. J Pharm Pharmacol 31(5):331–332

Cristian C, Karol P (2009) Encyclopedia of nanoscience and nanotechnology, 2nd edn. CRC Press, Florida

Cui X, Zhong S, Wang H (2007) Emulsifier-free core–shell polyacrylate latex nanoparticles containing fluorine and silicon in shell. Polymer 48(25):7241–7248. doi:10.1016/j.polymer.2007.10.019

Danielsson I, Lindman B (1981) The definition of microemulsion. Colloids Surf 3(4):391–392. doi:10.1016/0166-6622(81)80064-9

Darwish IA, Uchegbu IF (1997) The evaluation of crown ether based niosomes as cation containing and cation sensitive drug delivery systems. Int J Pharm 159(2):207–213. doi:10.1016/s0378-5173(97)00289-5

El-Shabouri M (2002) Positively charged nanoparticles for improving the oral bioavailability of cyclosporin-A. Int J Pharm 249(1–2):101–108. doi:10.1016/s0378-5173(02)00461-1

Fang J-Y, Leu Y-L, Chang C–C, Lin C-H, Tsai Y-H (2004) Lipid nano/submicron emulsions as vehicles for topical flurbiprofen delivery. Drug Delivery 11(2):97–105. doi:10.1080/10717540490280697

Fessi H, Puisieux F, Devissaguet JP, Ammoury N, Benita S (1989) Nanocapsule formation by interfacial polymer deposition following solvent displacement. Int J Pharm 55(1):R1–R4. doi:10.1016/0378-5173(89)90281-0

Forssen E, Willis M (1998) Ligand-targeted liposomes. Adv Drug Deliv Rev 29(3):249–271. doi:10.1016/s0169-409x(97)00083-5

Fukushima S, Kishimoto S, Takeuchi Y, Fukushima M (2000) Preparation and evaluation of o/w type emulsions containing antitumor prostaglandin. Adv Drug Deliv Rev 45(1):65–75. doi:10.1016/s0169-409x(00)00101-0

Gabizon AA (1992) Selective tumor localization and improved therapeutic index of anthracyclines encapsulated in long-circulating liposomes. Cancer Res 52(4):891–896

Gasco MR, Trotta M (1986) Nanoparticles from microemulsions. Int J Pharm 29(2–3):267–268

Gaudin F, Sintes-Zydowicz N (2008) Core–shell biocompatible polyurethane nanocapsules obtained by interfacial step polymerisation in miniemulsion. Colloids Surf A: Physicochem Eng Aspects 331(1–2):133–142. doi:10.1016/j.colsurfa.2008.07.028

Gianasi E, Cociancich F, Uchegbu IF, Florence AT, Duncan R (1997) Pharmaceutical and biological characterisation of a doxorubicin-polymer conjugate (PK1) entrapped in sorbitan monostearate Span 60 niosomes. Int J Pharm 148(2):139–148. doi:10.1016/s0378-5173(96)04840-5

Gref R, Minamitake Y, Peracchia M, Trubetskoy V, Torchilin V, Langer R (1994) Biodegradable long-circulating polymeric nanospheres. Science 263(5153):1600–1603. doi:10.1126/science.8128245

Gref R, Minamitake Y, Peracchia MT, Domb A, Trubetskoy V, Torchilin V, Langer R (1997) Poly(ethylene glycol)-coated nanospheres: potential carriers for intravenous drug administration. Pharm Biotechnol 10:167–198

Gurny R, Peppas NA, Harrington DD, Banker GS (1981) Development of biodegradable and injectable latices for controlled release of potent drugs. Drug Dev Ind Pharm 7(1):1–25. doi:10.3109/03639048109055684

Hawker CJ, Frechet JMJ (1990) Preparation of polymers with controlled molecular architecture. A new convergent approach to dendritic macromolecules. J Am Chem Soc 112(21):7638–7647. doi:10.1021/ja00177a027

Hecht S, Fréchet JMJ (2001) Dendritic encapsulation of function: applying nature's site isolation principle from biomimetics to materials science. Angew Chem Int Ed 40(1):74–91. doi:10.1002/1521-3773(20010105)40:1<74:aid-anie74>3.0.co;2-c

Hood E, Gonzalez M, Plaas A, Strom J, VanAuker M (2007) Immuno-targeting of nonionic surfactant vesicles to inflammation. Int J Pharm 339(1–2):222–230. doi:10.1016/j.ijpharm.2006.12.048

Hornig S, Heinze T (2007) Nanoscale structures of dextran esters. Carbohydr Polym 68(2):280–286. doi:10.1016/j.carbpol.2006.12.007

ICH-Guideline (2011) Impurities: guideline for residual solvents Q3C(R5). International conference on harmonisation of technical requirements for registration of pharmaceuticals for human use. http://www.ich.org/products/guidelines/quality/quality-single/article/impurities-guideline-for-residual-solvents.html. Accessed 10 May 2013

Jaiswal J, Kumar Gupta S, Kreuter J (2004) Preparation of biodegradable cyclosporine nanoparticles by high-pressure emulsification-solvent evaporation process. J Control Release 96(1):169–178. doi:10.1016/j.jconrel.2004.01.017

Janes KA, Calvo P, Alonso MJ (2001) Polysaccharide colloidal particles as delivery systems for macromolecules. Adv Drug Deliv Rev 47(1):83–97. doi:10.1016/S0169-409X(00)00123-X

Jansen JFGA, Meijer EW, de Brabander-van den Berg EMM (1995) The dendritic box: shape-selective liberation of encapsulated guests. J Am Chem Soc 117(15):4417–4418. doi:10.1021/ja00120a032

Jia G, Wang H, Yan L, Wang X, Pei R, Yan T, Zhao Y, Guo X (2005) Cytotoxicity of carbon nanomaterials: single-wall nanotube, multi-wall nanotube, and fullerene. Environ Sci Technol 39(5):1378–1383. doi:10.1021/es0487291

Kam NWS, Dai H (2005) Carbon nanotubes as intracellular protein transporters: generality and biological functionality. J Am Chem Soc 127(16):6021–6026. doi:10.1021/ja050062v

Karode SK, Kulkarni SS, Suresh AK, Mashelkar RA (1998) New insights into kinetics and thermodynamics of interfacial polymerization. Chem Eng Sci 53(15):2649–2663. doi:10.1016/s0009-2509(98)00083-9

Koosha F, Muller RH, Davis SS, Davies MC (1989) The surface chemical structure of poly(β-hydroxybutyrate) microparticles produced by solvent evaporation process. J Control Release 9(2):149–157. doi:10.1016/0168-3659(89)90005-9

Kreuter J, Speiser PP (1976) New adjuvants on a polymethylmethacrylate base. Infect Immun 13(1):204–210

Lamprecht A (2009) Nanotherapeutics: drug delivery concepts in nanoscience. Pan Stanford Publishing

Legrand P, Barratt G, Mosqueira V, Fessi H, Devissaguet JP (1999) Polymeric nanocapsules as drug delivery systems, a review. STP Pharma Sci 9(5):411–418

Leiza JR, Sudol ED, El-Aasser MS (1997) Preparation of high solids content poly(n-butyl acrylate) latexes through miniemulsion polymerization. J Appl Polym Sci 64(9):1797–1809. doi:10.1002/(sici)1097-4628(19970531)64:9<1797:aid-app16>3.0.co;2-v

Lesieur S, Grabielle-Madelmont C, Paternostre M-T, Moreau J-M, Handjani-Vila R-M, Ollivon M (1990) Action of octylglucoside on non-ionic monoalkyl amphiphile-cholesterol vesicles: study of the solubilization mechanism. Chem Phys Lipids 56(2–3):109–121. doi:10.1016/0009-3084(90)90094-8

Letchford K, Burt H (2007) A review of the formation and classification of amphiphilic block copolymer nanoparticulate structures: micelles, nanospheres, nanocapsules and polymersomes. Eur J Pharm Biopharm 65(3):259–269. doi:10.1016/j.ejpb.2006.11.009

Liu M, Kono K, Fréchet JMJ (2000) Water-soluble dendritic unimolecular micelles: their potential as drug delivery agents. J Control Release 65(1–2):121–131. doi:10.1016/s0168-3659(99)00245-x

Liu M, Zhou Z, Wang X, Xu J, Yang K, Cui Q, Chen X, Cao M, Weng J, Zhang Q (2007) Formation of poly(l, d-lactide) spheres with controlled size by direct dialysis. Polymer 48(19):5767–5779. doi:10.1016/j.polymer.2007.07.053

Liu Z, Jiao Y, Wang Y, Zhou C, Zhang Z (2008) Polysaccharides-based nanoparticles as drug delivery systems. Adv Drug Deliv Rev 60(15):1650–1662. doi:10.1016/j.addr.2008.09.001

Loup C, Zanta M-A, Caminade A-M, Majoral J-P, Meunier B (1999) Preparation of water-soluble cationic phosphorus-containing dendrimers as DNA transfecting agents. Chem: A Eur J 5(12):3644–3650. doi:10.1002/(sici)1521-3765(19991203)5:12<3644:aid-chem3644>3.0.co;2-i

Macías ER, Rodríguez-Guadarrama LA, Cisneros BA, Castañeda A, Mendizábal E, Puig JE (1995) Microemulsion polymerization of methyl methacrylate with the functional monomer N-methylolacrylamide. Colloids and Surf A: Physicochem Eng Aspects 103(1–2):119–126. doi:10.1016/0927-7757(95)03209-v

Mammen M, Choi S-K, Whitesides GM (1998) Polyvalent interactions in biological systems: implications for design and use of multivalent ligands and inhibitors. Angew Chem Int Ed 37(20):2754–2794.   doi:10.1002/(sici)1521-3773(19981102)37:20<2754:aid-anie2754>3.0.co;2-3

Martin TJ, Procházka K, Munk P, Webber SE (1996) pH-dependent micellization of poly(2-vinylpyridine)-block-poly(ethylene oxide). Macromolecules 29(18):6071–6073. doi:10.1021/ma960629f

Marty JJ, Oppenheim RC, Speiser P (1978) Nanoparticles-a new colloidal drug delivery system. Pharm Acta Helv 53(1):17–23

Mehnert W, Mäder K (2001) Solid lipid nanoparticles: production, characterization and applications. Adv Drug Deliv Rev 47(2–3):165–196. doi:10.1016/s0169-409x(01)00105-3

Meng F, Engbers GHM, Feijen J (2005) Biodegradable polymersomes as a basis for artificial cells: encapsulation, release and targeting. J Control Release 101(1–3):187–198. doi:10.1016/j.jconrel.2004.09.026

Meziani MJ, Pathak P, Hurezeanu R, Thies MC, Enick RM, Sun Y-P (2004) Supercritical-fluid processing technique for nanoscale polymer particles. Angewandte Chemie Int Ed 43(6):704–707. doi:10.1002/anie.200352834

Mitsukami Y, Donovan MS, Lowe AB, McCormick CL (2001) Water-soluble polymers. 81. Direct synthesis of hydrophilic styrenic-based homopolymers and block copolymers in aqueous solution via RAFT. Macromolecules 34(7):2248–2256. doi:10.1021/ma0018087

Molpeceres J, Guzman M, Aberturas MR, Chacon M, Berges L (1996) Application of central composite designs to the preparation of polycaprolactone nanoparticles by solvent displacement. J Pharm Sci 85(2):206–213. doi:10.1021/js950164r

Mouran D, Reimers J, Schork FJ (1996) Miniemulsion polymerization of methyl methacrylate with dodecyl mercaptan as cosurfactant. J Polym Sci Part A: Polym Chem 34(6):1073–1081. doi:10.1002/(sici)1099-0518(19960430)34:6<1073:aid-pola16>3.0.co;2-4

Muller RH, Mader K, Gohla S (2000) Solid lipid nanoparticles (SLN) for controlled drug delivery: a review of the state of the art. Eur J Pharm Biopharm 50(1):161–177. doi:10.1016/S0939-6411(00)00087-4

Müller RH, Radtke M, Wissing SA (2002) Nanostructured lipid matrices for improved microencapsulation of drugs. Int J Pharm 242(1–2):121–128. doi:10.1016/s0378-5173(02)00180-1

Muñoz-Bonilla A, van Herk AM, Heuts JPA (2010) Preparation of hairy particles and antifouling films using brush-type amphiphilic block copolymer surfactants in emulsion polymerization. Macromolecules 43(6):2721–2731. doi:10.1021/ma9027257

Needham D, Dewhirst MW (2001) The development and testing of a new temperature-sensitive drug delivery system for the treatment of solid tumors. Adv Drug Deliv Rev 53(3):285–305. doi:10.1016/s0169-409x(01)00233-2

Némati F, Dubernet C, Fessi H, Colin de Verdière A, Poupon MF, Puisieux F, Couvreur P (1996) Reversion of multidrug resistance using nanoparticles in vitro: Influence of the nature of the polymer. Int J Pharm 138(2):237–246. doi:10.1016/0378-5173(96)04559-0

Nicolaos G, Crauste-Manciet S, Farinotti R, Brossard D (2003) Improvement of cefpodoxime proxetil oral absorption in rats by an oil-in-water submicron emulsion. Int J Pharm 263(1–2):165–171. doi:10.1016/s0378-5173(03)00365-x

Pan G, Shawer M, Øie S, Lu DR (2003) In vitro gene transfection in human glioma cells using a novel and less cytotoxic artificial lipoprotein delivery system. Pharm Res 20(5):738–744. doi:10.1023/a:1023477317668

Peppas NA, Hilt JZ, Khademhosseini A, Langer R (2006) Hydrogels in biology and medicine: from molecular principles to bionanotechnology. Adv Mater 18(11):1345–1360. doi:10.1002/adma.200501612

Perez C, Sanchez A, Putnam D, Ting D, Langer R, Alonso MJ (2001) Poly(lactic acid)-poly(ethylene glycol) nanoparticles as new carriers for the delivery of plasmid DNA. J Control Release 75(1–2):211–224. doi:10.1016/s0168-3659(01)00397-2

Pietkiewicz J, Sznitowska M, Placzek M (2006) The expulsion of lipophilic drugs from the cores of solid lipid microspheres in diluted suspensions and in concentrates. Int J Pharm 310(1–2):64–71. doi:10.1016/j.ijpharm.2005.11.038

Pinto Reis C, Neufeld RJ, Ribeiro AJ, Veiga F (2006) Nanoencapsulation I. Methods for preparation of drug-loaded polymeric nanoparticles. Nanomed Nanotechnol Biol Med 2(1):8–21. doi:10.1016/j.nano.2005.12.003

Polidori A, Pucci B, Riess JG, Zarif L, Pavia AA (1994) Synthesis of double-chain glycolipids derived from aspartic acid: preliminary investigation of their colloidal behavior. Tetrahedron Lett 35(18):2899–2902. doi:10.1016/s0040-4039(00)76654-8

Prabaharan M, Mano JF (2005) Chitosan-based particles as controlled drug delivery systems. Drug Deliv 12(1):41–57

Quintanar-Guerrero D, Allemann E, Doelker E, Fessi H (1998) Preparation and characterization of nanocapsules from preformed polymers by a new process based on emulsification-diffusion technique. Pharm Res 15(7):1056–1062

Quintanar-Guerrero D, Allemann E, Fessi H, Doelker E (1999) Pseudolatex preparation using a novel emulsion-diffusion process involving direct displacement of partially water-miscible solvents by distillation. Int J Pharm 188(2):155–164

Rabinovich-Guilatt L, Couvreur P, Lambert G, Dubernet C (2004) Cationic vectors in ocular drug delivery. J Drug Target 12(9–10):623–633. doi:10.1080/10611860400015910

Rao JP, Geckeler KE (2011) Polymer nanoparticles: preparation techniques and size-control parameters. Prog Polym Sci 36(7):887–913. doi:10.1016/j.progpolymsci.2011.01.001

Rijcken CJF, Soga O, Hennink WE, Nostrum CV (2007) Triggered destabilisation of polymeric micelles and vesicles by changing polymers polarity: an attractive tool for drug delivery. J Control Release 120(3):131–148. doi:10.1016/j.jconrel.2007.03.023

Samad A, Sultana Y, Aqil M (2007) Liposomal drug delivery systems: an update review. Curr Drug Deliv 4(4):297–305

Sane A, Thies MC (2007) Effect of material properties and processing conditions on RESS of poly(l-lactide). J Supercrit Fluids 40(1):134–143. doi:10.1016/j.supflu.2006.04.003

Scheffel U, Rhodes BA, Natarajan TK, Wagner HN Jr (1972) Albumin microspheres for study of the reticuloendothelial system. J Nucl Med 13(7):498–503

Selvi BR, Jagadeesan D, Suma BS, Nagashankar G, Arif M, Balasubramanyam K, Eswara-moorthy M, Kundu TK (2008) Intrinsically fluorescent carbon nanospheres as a nuclear targeting vector: delivery of membrane-impermeable molecule to modulate gene expression in vivo. Nano Lett 8(10):3182–3188. doi:10.1021/nl801503m

Sheikh F, Barakat N, Kanjwal M, Aryal S, Khil M, Kim H-Y (2009) Novel self-assembled amphiphilic poly(ε-caprolactone)-grafted-poly(vinyl alcohol) nanoparticles: hydrophobic and hydrophilic drugs carrier nanoparticles. J Mater Sci Mater Med 20(3):821–831. doi:10.1007/s10856-008-3637-5

Solans C, Izquierdo P, Nolla J, Azemar N, Garcia-Celma MJ (2005) Nano-emulsions. Curr Opin Colloid Interface Sci 10(3–4):102–110. doi:10.1016/j.cocis.2005.06.004

Song X, Zhao Y, Wu W, Bi Y, Cai Z, Chen Q, Li Y, Hou S (2008) PLGA nanoparticles simultaneously loaded with vincristine sulfate and verapamil hydrochloride: systematic study of particle size and drug entrapment efficiency. Int J Pharm 350(1–2):320–329. doi:10.1016/j.ijpharm.2007.08.034

Sosa N, Zaragoza EA, López RG, Peralta RD, Katime I, Becerra F, Mendizábal E, Puig JE (2000) Unusual free radical polymerization of vinyl acetate in anionic microemulsion media. Langmuir 16(8):3612–3619. doi:10.1021/la991065m

Soussan E, Cassel S, Blanzat M, Rico-Lattes I (2009) Drug delivery by soft matter: matrix and vesicular carriers. Angew Chem Int Ed Engl 48(2):274–288. doi:10.1002/anie.200802453

Swarbrick J (2006) Encyclopedia of pharmaceutical technology. Informa Healthcare, New York

Tabata Y, Ikada Y (1989) Protein precoating of polylactide microspheres containing a lipophilic immunopotentiator for enhancement of macrophage phagocytosis and activation. Pharm Res 6(4):296–301. doi:10.1023/a:1015942306801

Tamilvanan S (2004) Oil-in-water lipid emulsions: implications for parenteral and ocular delivering systems. Prog Lipid Res 43(6):489–533

Tamilvanan S, Schmidt S, Müller RH, Benita S (2005) In vitro adsorption of plasma proteins onto the surface (charges) modified-submicron emulsions for intravenous administration. Eur J Pharm Biopharm 59(1):1–7. doi:10.1016/j.ejpb.2004.07.001

Tang MX, Szoka FC (1997) The influence of polymer structure on the interactions of cationic polymers with DNA and morphology of the resulting complexes. Gene Ther 4(8):823–832. doi:10.1038/sj.gt.3300454

Tang MX, Redemann CT, Szoka FC (1996) In vitrogene delivery by degraded polyamidoamine dendrimers. Bioconjug Chem 7(6):703–714. doi:10.1021/bc9600630

Teli KM, Mutalik S, Rajanikant GK (2010) Nanotechnology and nanomedicine: going small means aiming big. Curr Pharm Des 16(16):1882–1892. doi:10.2174/138161210791208992

Tice TR, Gilley RM (1985) Preparation of injectable controlled-release microcapsules by a solvent-evaporation process. J Control Release 2:343–352. doi:10.1016/0168-3659(85)90056-2

Tobío M, Gref R, Sánchez A, Langer R, Alonso MJ (1998) Stealth PLA-PEG nanoparticles as protein carriers for nasal administration. Pharm Res 15(2):270–275. doi:10.1023/a:1011922819926

Tomalia DA, Baker H, Dewald J, Hall M, Kallos G, Martin S, Roeck J, Ryder J, Smith P (1985) A new class of polymers: starburst-dendritic macromolecules. Polym J 17(1):117–132

Uchegbu IF, Vyas SP (1998) Non-ionic surfactant based vesicles (niosomes) in drug delivery. Int J Pharm 172(1–2):33–70. doi:10.1016/s0378-5173(98)00169-0

Ueda M, Kreuter J (1997) Optimization of the preparation of loperamide-loaded poly (L-lactide) nanoparticles by high pressure emulsification-solvent evaporation. J Microencapsul 14(5):593–605. doi:10.3109/02652049709006812

Westesen K, Siekmann B, Koch MHJ (1993) Investigations on the physical state of lipid nanoparticles by synchrotron radiation X-ray diffraction. Int J Pharm 93(1–3):189–199. doi:10.1016/0378-5173(93)90177-h

Wissing SA, Kayser O, Müller RH (2004) Solid lipid nanoparticles for parenteral drug delivery. Adv Drug Deliv Rev 56(9):1257–1272. doi:10.1016/j.addr.2003.12.002

Wu H, Ramachandran C, Bielinska AU, Kingzett K, Sun R, Weiner ND, Roessler BJ (2001) Topical transfection using plasmid DNA in a water-in-oil nanoemulsion. Int J Pharm 221(1–2):23–34. doi:10.1016/s0378-5173(01)00672-x

Yoo HS, Oh JE, Lee KH, Park TG (1999) Biodegradable nanoparticles containing doxorubicin-PLGA conjugate for sustained release. Pharm Res 16(7):1114–1118

Yuk SH, Cho SH, Lee SH (1997) pH/temperature-responsive polymer composed of poly((N, N-dimethylamino)ethyl methacrylate-co-ethylacrylamide). Macromolecules 30(22):6856–6859. doi:10.1021/ma970725w

Zambaux MF, Bonneaux F, Gref R, Maincent P, Dellacherie E, Alonso MJ, Labrude P, Vigneron C (1998) Influence of experimental parameters on the characteristics of poly(lactic acid) nanoparticles prepared by a double emulsion method. J Control Release 50(1–3):31–40

Zinselmeyer BH, Mackay SP, Schatzlein AG, Uchegbu IF (2002) The lower-generation polypropylenimine dendrimers are effective gene-transfer agents. Pharm Res 19(7):960–967

Chapter 11
Nanoparticulate Formulations for Pesticide Applications

The following sections give details of the nanoparticulate formulation developed by various researchers. The innovative technology to formulate the nanoparticle is briefly discussed along the entrapment strategies. Few biological models for testing the efficacy of these developed formulations on insect model have also been included.

11.1 Microemulsion

Microemulsion is an isotropic mixture of at least a hydrophilic, a hydrophobic, and an amphiphilic component. Their thermodynamic stability and their nanostructure are two important characteristics that distinguish them from ordinary emulsions which are thermodynamically unstable. The term "microemulsion" was first coined by Schulman et al. (1959). Schulman et al. defined the microemulsion as "dispersions consist of uniform spherical droplets of either oil or water dispersed in the appropriate continuous phase."

Construction of phase diagram is an essential tool for designing a microemulsion system. The primary aim of phase diagram is to find the conditions under which the surfactant can solubilize the maximum amount of water/ oil. Construction of phase diagram enables determination of aqueous dilutability and range of compositions that can form a monophasic region (Fig. 11.1).

Despite their name, microemulsions are fundamentally different from emulsions and should not be seen as mere emulsions with a small droplet size. Microemulsions are thermodynamically stable systems and display indefinite stability in the absence of chemical degradation of any of its components. Emulsions, on the other hand, are merely kinetically stabilized but thermodynamically unstable, which means that emulsions will eventually separate to macroscopically separated oil and water phases. Due to their thermodynamic stability, microemulsions form spontaneously and no work has to be added to prepare them. Emulsions, on the other hand, are not thermodynamically stable; and hence, energy must be added to form them. Emulsions consist of relatively large droplets, whereas microemulsions

A. De et al., *Targeted Delivery of Pesticides Using Biodegradable Polymeric Nanoparticles*, SpringerBriefs in Molecular Science,
DOI: 10.1007/978-81-322-1689-6_11, © The Author(s) 2014

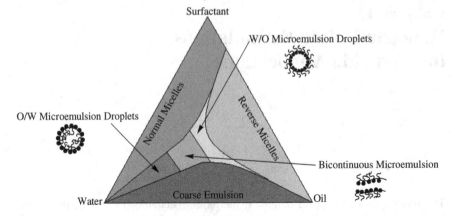

Fig. 11.1 Hypothetical pseudo-ternary phase diagram of an oil/ surfactant/ water system showing regions of microemulsions and emulsion phases (Rane and Anderson 2008; Lawrence and Rees 2000; Prince 1975)

consist of very small droplets (typically 10–100 nm). Microemulsion can also display a range of other structures. As a result of this, microemulsions are transparent, whereas emulsions are milky in their appearance. Due to the larger droplets in the emulsion systems, the surface area is generally smaller in emulsions than in microemulsions, and consequently, less surfactant is generally needed to generate an emulsion than a microemulsion system (Malmsten 2002). Moreover, microemulsions have very low surface tension and small droplet size. Typically, they are clear, transparent, thermodynamically stable dispersion of oil and water, stabilized by interfacial film of surfactant frequently in combination with a cosurfactant (Lawrence and Rees 2000).

Since emulsification is a non-spontaneous process that tends to be an unstable system, therefore, the pesticide emulsions are being increasingly replaced by microemulsion formulations. Microemulsions are formed spontaneously when the components are taken in an appropriate proportion, and therefore, the preparation requires less energy (Pratap and Bhowmick 2008). Microemulsions have been widely studied to enhance the bioavailability of the poorly soluble drugs.

Microemulsion offers a cost-effective approach for solubilizing poorly soluble active ingredients. It is to be noted that because they have very low surface tension and small droplet size, high absorption and permeation can be usually achieved (Talegaonkar et al. 2008).

11.1.1 Surfactants in Microemulsion

Surfactants can perform several functions in the design of formulations (Pratap and Bhowmick 2008). Both ionic and non-ionic types of surfactants can be used for the preparation of microemulsions. Formation of micelles is a consequence of the fact that surfactants are curvature loving. Thickness of a typical surfactant film is of the order of 1 nm. Good emulsifying agents prefer to form monolayers of low curvature and their head group and tail repulsions nearly balance each other. Nowadays, the preference is toward the use of non-ionics in order to optimize the inertness between the surfactant and the entrapped molecule and to minimize the effect of varying water hardness. They primarily reduce the surface tension and consequently, in the case of agrochemicals, increase the spreadability of the drop and adhesion wetting on the surface. This helps in reducing the rate of evaporation. Surfactants were also found to modify the phytotoxicity of pesticides.

11.1.2 Role of Cosurfactant

Cosurfactants are normally used in combination with surfactants so as to improve their properties (Pratap and Bhowmick 2008). These cosurfactants can be ionic substances having a strong polarity or hydrophobic chains (which can be more or less long and branched) and attached to a hydrophilic site such as $-CH_2OH$, $-CHOH$, $-COOH$. Cosurfactant molecules, typically, do not form micelles but can stay in combined micelle structure (for example, n-butanol, n-octanol, n-octylamine) and bring about a synergistic effect more drastically. Surfactant substances along with cosurfactants can modify the emulsion structure by lowering the interfacial tension within the required limits. Basically, cosurfactants can partly screen the electrostatic repulsion between the charged head groups of the surfactants and act as a spacer. Short cosurfactants chains can also reduce the interfacial repulsion, thus lowering the bending rigidity of the interfacial film.

Anjali el al. (2010) have prepared a microemulsion formulation by using the pesticide permethrin for larvicidal application. Permentrin (Fig. 11.2) is a viscous liquid at the room temperature. It does not dissociate in water and has extremely low solubility, with saturation solubility of 11.1 µg/mL (100% purity) at 20 °C. Microemulsion was prepared by dissolving permethrin in an organic solvent n-butyl acetate. To this ammonium glycyrrhizinate (Fig. 11.3), sec-butyl alcohol and soybean lecithin with soybean phosphatidylcholine were added to create an oil phase. The above oil phase was emulsified with an aqueous solution of sucrose.

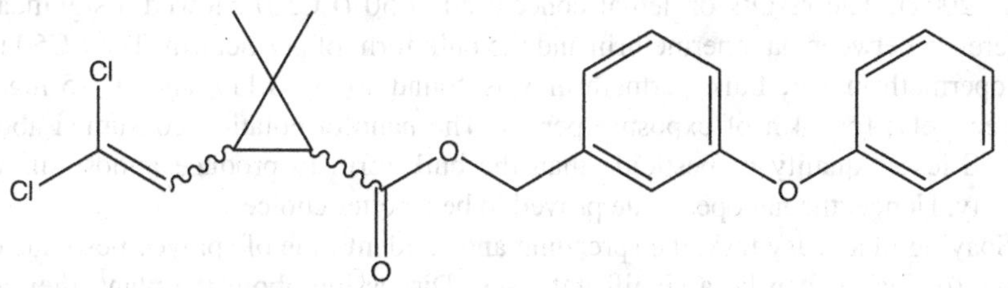

Fig. 11.2 Chemical structure of permethrin

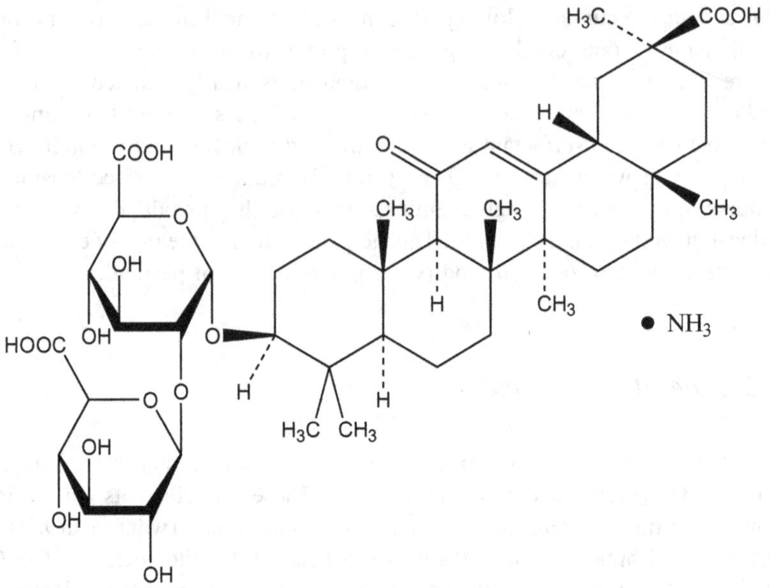

Fig. 11.3 Chemical structure of ammonium glycyrrhizinate (European-Pharmacopoeia 2005)

The mixture was allowed to equilibrate till an isotropic system was formed. Furthermore, the microemulsion was dried by lyophilization. The dried power was easily dispersible in water that yielded a translucent and stable dispersion.

The permethrin nanoparticles were amorphous and the average particle size was found to be 151 nm. The bioassay of the microemulsion formulation was performed on *Culex quinquefasciatus*, a vector of lymphatic filariasis, which is a tropical disease. Around 120 million people are infected worldwide; forty-four million have chronic manifestation. Though there are lots of larvicidal agents against mosquitoes, all suffers from poor water solubility, which is a limiting factor in the development of the suitable formulations. When the particle size of the poorly water-soluble larvicidal agent is reduced to a nanoscale, it results in higher water solubility and dissolution rate (Patravale and Kulkarni 2004). Downsizing the pesticide also induces systemic activity due to the higher mobility of the particles (Anees 2008). The resistance of *Cx. quinquefasciatus* larvae toward permethrin is 13–940-folds (Liu et al. 2004). The results on lethal concentration 50 (LC50) showed a significant difference between nanopermethrin and the bulk form of permethrin. The LC50 for nanopermethrin and bulk permethrin was found to be 0.117 and 0.715 mg/L, respectively, for 24 h of exposure period. The nanoformulation consumed about sixfold lesser quality of pesticide than the bulk form to produce almost similar activity. Hence, the nanopesticide proved to be a better choice.

Spaying in an easy task, the spreading and resident time of sprayed pesticide on plants (or insect) can be a significant issue. Discussing about the plant, they are naturally covered by trichromes and cuticles which is an essential part of their self-defense system. The plant cuticle is a protective waxy film covering the epidermis of leaves, shoots, and other aerial parts of plant. The cuticular membrane is impregnated with cuticular waxes (which are mixture of hydrophobic aliphatic compounds with hydrocarbon chain length ranging from C-16 to C-17) (Hemsley and Poole 2004). The waxy covering of the cuticle on plant epidermis produces a water impermeable layer/film which imparts a problem for pesticide spays formulations. Spraying involves delivery of liquid formulation at high impact velocity at the aerial parts of the plants. The dynamic behavior of the droplet after impingement depends on the various factors, viz. surface tension and viscosity of the solution; droplet size and velocity of the spray; surface texture or toughness of the solid target. The impact of liquid droplets after hitting solid surfaces may cause spreading, recoil, and splashing of the droplets (Lee and Lee 2011; Vadillo et al. 2009). This process gives very less time to deposit the pesticide formulation on the plant surface protected by waxy cuticle layer/ film. These factors can affect the wettability and resident time of the droplet on leafy surfaces. In order for pesticides to have a high efficiency, they must have ability to spread on and adhere to plant surfaces. Failing to fulfill such demands can lead to repeated application of spraying the formulation on the crop (and this has becomes necessary routine procedure by the farmers). Naturally, the success of such liquid formulation depends on their ability for wetting and spreading themselves on the plant surfaces. An aqueous-based cypermethrin microemulsion was prepared by Zhang et al. (2013) by adding oil to emulsified water, with ethyl butyrate as the solvent and TritonX-100 (TX-100) (Fig. 11.4a) and sodium dodecyl benzene sulfonate (Fig. 11.4b) as surfactants together with and n-butyl alcohol (n-C_4H_9OH) as a cosurfactant. Youngfu wheat plant was selected as model plant in this study. The microemulsion showed a low contact angle and low surface tension. Moreover, the droplet radius was about 45 nm.

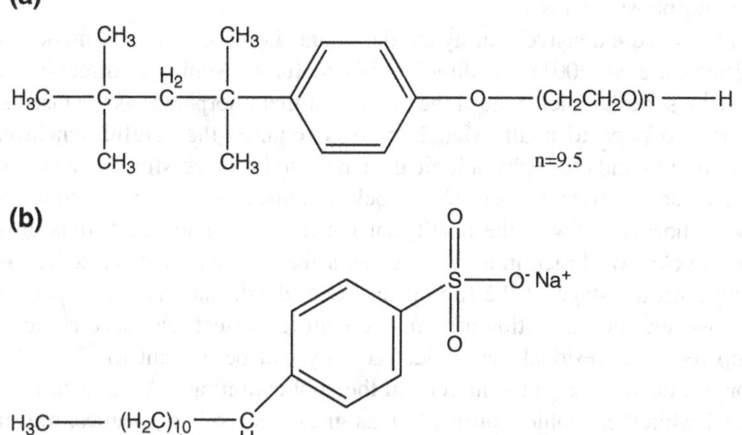

(a)

(b)

Fig. 11.4 Chemical structure of **a** Triton X-100 and **b** sodium dodecyl benzene sulfonate (Sigma Chemical; Sigma-Aldrich)

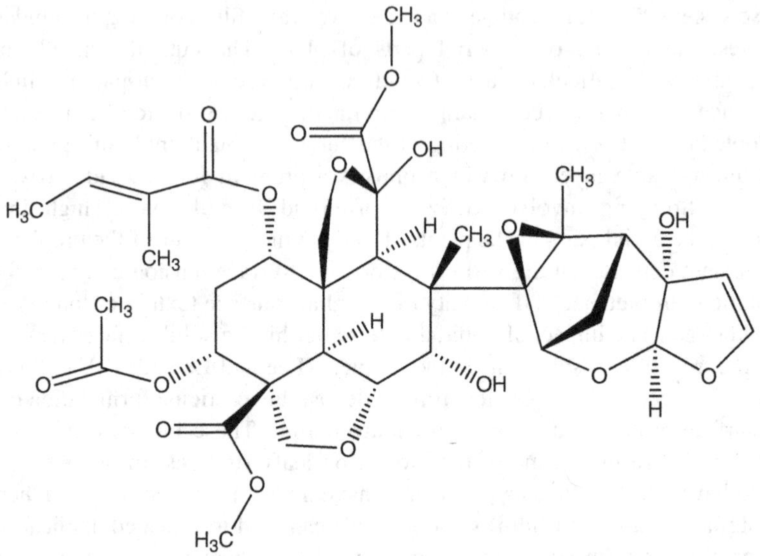

Fig. 11.5 Chemical structure of azadirachtin (Nisbet 2000)

Singla and Patanjali have prepared the microemulsion for herbal pesticide—neem oil (Singla and Patanjali 2013). The main challenge with oil-based formulation is that the concentrates get destroyed upon dilution with the aqueous phase and also lead to migration of the solubilized guest active molecule to the outer continuous phase, followed by precipitation and uncontrolled absorption. The Indian neem tree (*Azadirachta indica*) has been known through the ages for its medicinal and insect-repellent properties. Azadirachtin (Fig. 11.5), a tetranortriterpenoid, is the major active ingredient isolated from neem and is known to disrupt the metamorphosis of insects.

Neem is used extensively in ayurveda, unani, homeopathy, and modern medicine (Raizada et al. 2001). Azadirachtin is structurally similar to insect hormones called "ecdysones" which control the process of metamorphosis as the insects pass from larva to pupa to adult. Metamorphosis requires the careful synchrony of many hormones and other physiological changes to be successful, and azadirachtin seems to be an "ecdysone blocker." It blocks the insect's production and release of these vital hormones, with the result that the insects cannot molt, thus breaking their life cycle. Azadirachtin also serves as a feeding deterrent for some insects. Depending on the stage of the life cycle, insect death may not occur for several days. However, upon ingestion of minute quantities, insects can become quiescent and stop feeding. Residual insecticidal activity can be evident for 7–10 days or ever longer, depending on the insect and the application rate. Azadirachtin is used to control whiteflies, aphids, thrips, fungus gnats, caterpillars, beetles, mushroom flies, mealybugs, leaf miners, gypsy moths, and others on food, greenhouse crops, ornamentals, and turf. Neem oil is a registered pesticide in USA and is regarded as a general-use pesticide with a toxicity classification of IV (relatively non-toxic) (EXTOXNET 1995). Moreover, EPA has registered neem oil as a biopesticide suggesting that the application of neem oil is safe to mammals, birds, fish and aquatic invertebrates, other non-target insects, or plants (USEPA-OPP 2009). The study included the novel idea of preparing a water-based formulation and replacing the petroleum-based solvent for agrochemical formulations. Despite the fact that petroleum-based solvents can create a number of problems in agrochemical formulations such as irritation to the eyes at the time of manufacturing and applications and phytotoxicity to the plants, these have been used traditionally for many years. Hence, neem oil as an active solubilizate was used with 371-N (a nonyl-phenol-ethoxylate-based non-ionic surfactant) and Tween-60 as surfactants, n-butanol as the cosurfactant, and water as the aqueous phase. Oil phase was initially mixed with surfactant/cosurfactant and vortexed vigorously while titrating with water. The formation of microemulsion was confirmed visually through cross-polarized light. All the diluted samples were found to be isotopic. The hydrodynamic droplet radii measured by DLS for the diluted microemulsion droplets ranged between 1.9 and 8.5 nm, which kept increasing as the dilution of the microemulsion was increased. The increase in the hydrodynamic droplet radii of the diluted microemulsion samples maybe because of decrease in volume fraction or because of attractive influence between the diluted microemulsion droplets. In diluted microemulsion systems, the increase in diffusion coefficient of the droplet may be largely attributed to the attractive interactions, and this ultimately may result in increase in the hydrodynamic radii of the diluted samples. The neem oil microemulsion formulation consumed only 17% of the surfactant to entrap 12% of the neem oil. Diluted samples of the formulated microemulsion remained clear, isotropic, and one-phase system, and this may promote their use as a delivery system for botanical pesticides in safe pest management. Since, the active ingredient was probably entrapped by the interfacial surface film, therefore, the photostability of the active ingredient was also enhanced. Besides, another important advantage of these systems is their environment friendliness and triggered release of active ingredients because of high interfacial area and low interfacial tension. In addition, this system has several advantages, viz. procurement of neem oil from neem seed (which is of course of natural origin and is absolutely pollution free), the O/W type microemulsion system eliminates the use of petroleum product and uses water as a vehicle, which is not only economic but also safe.

It has been observed that the synthesis of proteinase inhibitors (PIs) has evolved naturally in plant to act as defense mechanisms against pests by interfering with their digestive biochemistry. The digestive system of the insects secretes a variety of enzymes that can breakdown carbohydrates and proteins presents in the ingested food. While doing so, they rapidly damage the crop. Plants often respond to this invasion by synthesizing proteins that can inhibit the action of the enzymes present in the insect gut, thereby depriving the digestion process which can lead to the insect starvation and growth retardation (Koundal and Rajendran 2003). Tamhane et al. (2012) have formulated a water-soluble plant protease inhibitor CanPI-7 as a biopesticide application. CanPI protein is synthesized by *Capsicum annuum* leaves and has 1–4 inhibitory domains with varied trypsin or chymotrypsin inhibitory sites. CanPIs are effective in reducing the fecundity of moths and fertility of eggs at low concentrations and can cause larval mortality at higher concentration (Tamhane et al. 2005). *Helicoverpa armigera* is one of the most devastating field pests for many important crops and can cause severe economic losses. Identification of specificities of PIs having high binding efficiency for insect gut enzymes was necessary for effective inhibition of midgut proteinases of *H. armigera*.

A bicontinuous microemulsion system was prepared for CanPI-7 for biopesticide delivery. CanPI-7 was dissolved in water and mixed with propanol and 1-butanol. The spreadability and wettability of the bicontinuous microemulsion system was estimated by determining the contact angle. The contact angle higher than 90° shows poor spreadability and wettability of the formulations. A drop was placed at the surface of chickpea leaves and it showed the value of contact angle at 26.7° which was significantly lower than the limit. This ensured greater spreadability and wettability on the leaf surface. The retention activity of the CanPI-7 was also evaluated. It has been observed that trypsin can digest the protein–gelatin. The capability of protease inhibitor to stop/ inhibit the digestion of gelatin was measured. Trypsin can digest gelatin on an unexposed X-ray film, thereby exposing the area of the film as a "blot" after washing (hence the name "dot blot assay"). Inhibition of the trypsin activity would leave the gelatin film undigested, and hence, no blot would be formed. However, exposure spots with trypsin in water or bicontinuous microemulsion (both containing CanPI-7) were left the film undigested. This clearly demonstrated that the protease inhibition activity of CanPI-7 was preserved when incorporated in the bicontinuous microemulsion and was comparable to its activity in water. In another experiment to assess the protease inhibitory activity, N_α-benzoyl-DL-arginine-p-nitroanilide hydrochloride (BApNA) was used as a substrate to measure the inhibition activity of CanPI-7 on trypsin and *H. armigera* gut proteases. The amide bond in BApNA is digested by proteases to yield N_α-benzoyl-DL-arginine and p-nitroaniline. The formation of p-nitroaniline, a yellow-colored product, would be proportional to the amount and activity of the protease present. Inhibition of the protease activity by CanPI-7 resulted in a decrease in the spectral absorbance of the product formed. It was observed that almost 90 % of trypsin inhibition (TI) and 60 % of the *Helicoverpa* gut protease inhibition (HGPI) could be obtained for CanPI-7 in both water and bicontinuous microemulsion. CanPI-7 incorporated in bicontinuous microemulsion was also tested for ex vivo leaf stability and activity. In this experiment, CanPI-7 lodged on the leaf surface using water or bicontinuous microemulsion was later recovered through water wash and quantified using the trypsin inhibition assay. Remarkably, inhibition activities measured from wash concentrate containing CanPI-7 recovered from bicontinuous microemulsion-spotted leaflets were three times less than the inhibition activity of the corresponding wash concentrate for the CanPI-7 recovered from water-spotted leaflets. The activities determined at 3- and 24-h time points did not have a significant difference indicating that the activity, and hence, the CanPI-7 retention on leaf surface, was not influenced by the duration of contact of the CanPI-7 with the leaf surface in either cases. These experiments clearly indicate that when CanPI-7 incorporated in bicontinuous microemulsion is placed on the leaf surface, permeation of CanPI-7 onto the leaf is taking place almost instantaneously and is further maintained (at least up to 24 h). Whereas with CanPI-7 aqueous solution, there was no detectable leaf permeation of CanPI-7 as the activity was almost completely recovered from the corresponding wash concentrate. Protecting plants against pests/pathogen using spray formulations with various chemicals is difficult and profligate if the weather is humid or it is raining. Simple water-based formulations do not wet the leaf surface efficiently and, therefore, do not spread evenly over the surface. Moreover, the moisture that condenses on the leaf surface during dusk and dawn easily washes out the sprayed protective ingredient necessitating repeated sprays. On the contrary, incorporation of the active ingredient in the plant through the use of bicontinuous microemulsion spray formulation will be beneficial providing continuous, longer-lasting protection to plants and reduce the frequency of spraying. Multiple PIs/ other defense molecules can also be solubilized and incorporated in the microemulsion-based delivery system and further lodged into the plant. Due to the high water content in the microemulsion, the bicontinuous microemulsion-based delivery is also a greener alternative to petroleum-based solvents widely used for pesticides. The volatile components in the bicontinuous microemulsion evaporate easily leaving behind no harmful residues. 1-Butanol and 2-propanol if taken up by humans/animals are known to follow the metabolic pathway similar to other alcohols. They undergo almost complete degradation by alcohol dehydrogenase to CO_2 and water. Thus, all the ingredients used in the bicontinuous microemulsion are eco-friendly and are comparatively more tolerable by the environment. Using bicontinuous microemulsion as a carrier for delivering the CanPI-7, which by its own virtue is an environmentally safe insect growth retarding protein sourced from a food crop, holds promise toward eco-friendly pest management. With the widely accessible recombinant approaches for producing protein, direct application of PI on plant surfaces is an efficient, eco-friendly, and practicable alternative to chemical pesticides. Moreover, PIs from edible sources and delivered as biopesticides have greater acceptability as they overcome the ethical considerations and other limitations posed by genetic modification approaches. However, efficient application of protein-based biopesticides to counter biotic stresses caused by insect pests involves several challenges that demand optimization and interfacing of surface chemistry solutions. Encapsulation and delivery of water-soluble PI onto hydrophobic leaf surfaces, hitherto unexplored, necessitate the use of an appropriate vehicle, especially in the form of a liquid formulation. The encapsulating media has to stabilize the protein, retain its activity, and finally enable adsorption as well as efficient surface permeation upon interaction with the surface of interest. In addition, the media itself should be stable under harsh environmental conditions and avoid any damage to the leaves or the plant growth in general. Bicontinuous microemulsion provides "greener" and hospitable environment for the encapsulation of protein while their interfacial properties enable efficient delivery.

The roots of *Derris elliptica* (Fabaceae) is the natural source of the insecticide rotenone (Li and Geng 2013). Several studies have confirmed its bioactive properties, such as insecticidal activity (Hu et al. 2005), resistance to plant viruses, and antitumor activity (Li and Geng 2013). Rotenoids have been used as crop insecticides since 1848, and they have been applied to plants so as to control leaf-eating caterpillars. However, the crude plant has also been used traditionally for fish poison (Metcalf 1948). Rotenone (Fig. 11.6) shows a pyrethrin-like behavior but with a stronger action and a higher persistence (Crombie 1999). It owes part of its efficacy to its rapid neurotoxic action against insects, named "knock-down effect." It is used to control aphids, suckers, thrips, and other insects on fruit and vegetables (Tomlin 2000). Rotenone not only increases the insect's mortality, but it also negatively affects its reproduction (Guadaño et al. 2000). However, there are only a few studies about the feeding-deterrent activity of rotenone and its derivatives (Bentley et al. 1987; Nawrot et al. 1989). When ingested, rotenone tends to reduce the amount of food absorbed by the larvae, as well as their ability to convert the absorbed food to biomass (Wheeler et al. 2001).

In another study, rotenone was formulated in a microemulsion system by using Edentor ME, Agnique BL 7001, and Agnique BL 7002 as oil (Ahmad et al. 2012) The oil phase was emulsified in aqueous phase in the presence of surfactant Tween 20, Agnique PG 8107-U, and Agnique PG 9116. The size of the microemulsion droplets ranged from 25 to 227 nm. The efficacy of the formulation was assayed by using leaf-dip-technique method. The first generation of early third instar Diamondback moth larvae, *Plutella xylostella* L. (Lepidotera: Plutellidae), was used as a pest model. Initially mustard leaf, *Brassica junsea* (Brassicales: Brassicaceae),

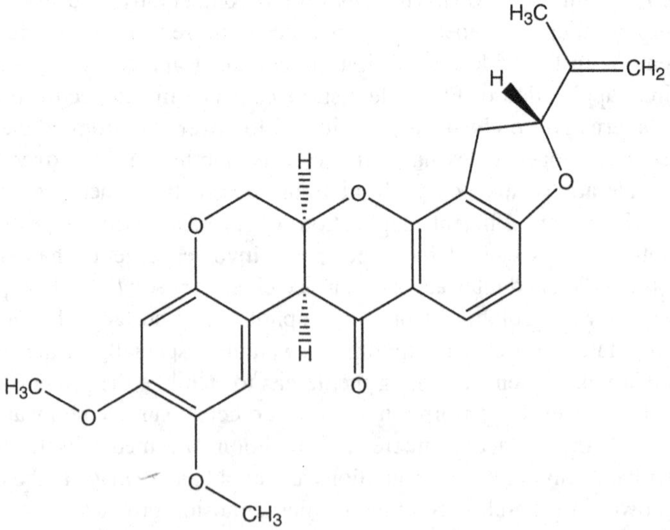

Fig. 11.6 Chemical structure of rotenone

disk (diameter 5.0 cm) was dipped in the microemulsion solution for 10 s and then the leaf was allowed to dry. The larvae were placed on the treated leaf, and mortality of the nymps were recorded after 96 h. The highest mortality (with lowest LC50) was observed for the formation synthesized by Agnique BL 7002/Agnique PG 8107-U/water combination. Agnique® PG are EPA-approved alkyl polyglycoside (C8-C16)-based surfactants which are exceptionally mild, readily biodegradable, and synthesized from renewable plant-based raw materials (Cognis 2007).

Pratap and Bhowmick have shown the application of acid oil in the delivery of chlorpyrifos (Fig. 11.7) (Pratap and Bhowmick 2008). Acid oil is a by-product of edible oil refinery. Acid oil was saponified to obtain the methyl ester, and this methyl ester of acid oil was used as a solvent for the pesticide. The investigators have proposed O/W and W/O types of microemulsion systems in which chlorpyrifos was dissolved the methyl ester of the acid oil. In the case of O/W type system, a surfactant combination Unitop 100 and Unitop FFT 40 (1:1v/v) was used and n-butanol was used as a cosurfactant. Unitop 100 is nonylphenol ethoxylated with 9.5 mol of ethylene oxide, and Unitop FFT 40 is castor oil ethoxylated with 40 mol of ethylene oxide. In the case of W/O type system, rest of the ingredients were same except the surfactant Unitop 100 was used along with Hydol 6 (lauryl alcohol ethoxylated with 6 mol of ethylene oxide). The cosurfactants brought down the interfacial tension synergistically in the presence of the surfactants. Therefore, it was ensured that the preferred combination must have maximum percentage of cosurfactant and minimum percentage of surfactant. It was observed that a maximum of 12.30 % of cosurfactant and a minimum of 10.76 % of surfactant mixture gave an economical O/W microemulsion formulation. Similarly, a maximum of 14.28 % of cosurfactant and minimum of 14.28 % of surfactant mixture gave an economical W/O microemulsion formulation. It was thus shown that the application of a vegetable oil (or a solvent based on the vegetable oil), which is biodegradable and renewable, could serve as an alternative carrier for pesticides.

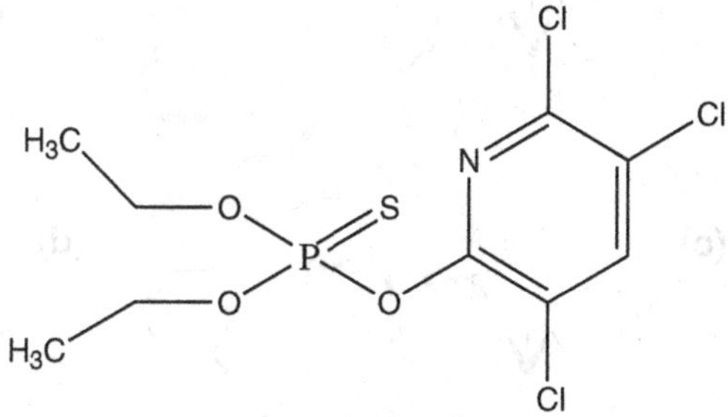

Fig. 11.7 Chemical structure of chlorpyrifos

11.1.3 Nanoemulsions

A low-energy self-emulsifying alcohol-free O/W nanoemulsions has been synthesized by Wang et al. (2007), by a two-step process. This involved crash dilution of a bicontinuous or oil-in-water microemulsion into a large volume of water at constant temperature (25 °C). The fatty acid methyl ester derivative, methyl decanoate, was used as oil phase. The fatty acid methyl esters are derived from natural source such as vegetable oils and have gained attention in past few years as solvents and other applications (Chhetri et al. 2008; Mohibbe Azam et al. 2005). Their economic viability has encouraged to explore them as solvent in pesticide formulation. Moreover, β-cypermethrin (Fig. 11.8) has several folds higher solubility in methyl decanoate (468 mg/mL) as compared to water (1.13×10^{-4} mg/mL). Initially, β-cypermethrin, solvent methyl decanoate, and surfactant poly(oxyethylene) lauryl ether were mixed at an appropriate ratio to form a concentrate. This concentrate was injected to very large volume of water, kept under gentle stirring to form a nanoemulsion of droplet size 30 nm. This two-step process for nanoemulsion synthesis was easy to scale up and consumed less energy making the technique promising from both environmental and economic points of view.

Citronella oil extracted from *Cymbopogon nardus* (citronella) has been used as natural mosquito repellents (Maia and Moore 2011). Citronella oil has been used alone or in combination with some other pesticides in the concentration range of 0.5–15 % (w/v) (Trongtokit et al. 2005; Fradin and Day 2002). The US Environmental Protection Agency (US EPA) has registered citronella oils as insect-repellent ingredients for application on the skin (Nerio et al. 2010). Citronella oil encapsulated nanoemulsion was prepared by a high-pressure homogenization technique. The transparent nanoemulsion of citronella oil (20 %) was obtained with 2.5 % of alkylpolyglucoside-based non-ionic surfactant (Montanov^{TM} 82) and 100 % glycerol in water. The mean size range for volatile-oil-loaded

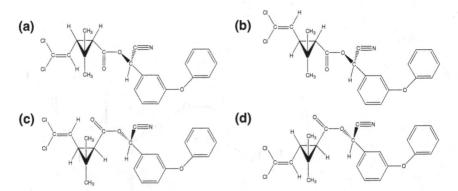

Fig. 11.8 Chemical structure of β-cypermethrin **a** (R)-alcohol(1S)-cis-acid; **b** (R)-alcohol(1S)-trans-acid; **c** (S)-alcohol(1R)-cis-acid; **d** (S)-alcohol(1R)-trans-acid (Wood)

nanoemulsion was found to be 120–200 nm. Montanov^{TM} 82 is a mixture of cetearyl alcohol and cocoyl glucoside and has been used for skincare, sun care, and other personal care cosmetic preparations as it gives a rich feel and spray to thick cream textures (Masmoudi et al. 2006). Glycerol behaves as a cosolvent and maintains transparency of the nanoemulsion. The high-pressure homogenizing process reduced the droplet size as well as the polydispersity of the oil droplets. The release of encapsulated citronella oil could be effectively controlled by changing the amount of surfactant and glycerol. It was found that by increasing the concentration of the surfactant, the size of the droplet could be reduced thereby increasing the homogeneity of the system, leading to extension of release and protection time. The higher amount of glycerol controlled the release of citronella oil from the nanoemulsion at a slower rate leading to a sustained release formulation (Sakulku et al. 2009).

Glyphosate [N-(phosphonomethyl) glycine] is a non-selective foliar-applied herbicide for post-emergence control of weeds (Tuffi Santos et al. 2007). Glyphosate acts by inhibiting the biosynthesis of aromatic amino acids via shikimate pathway through deactivation of 5-enolpyruvyl shikimate-3-phosphate synthase (Amrhein et al. 1980; Steinrücken and Amrhein 1980). This herbicide has desirable environmental characteristics due to its rapid degradation and deactivation in soil.

Glyphosate isopropylamine (Fig. 11.9) is composed of an anionic site of carboxylate $(OH)_2POCH_2NHCH_2COO^-$ and cationic site of isopropylamine with low octanol–water partition coefficient (Kow). Owing to its low hydrophobicity, the penetration through hydrophobic epicuticular wax and cuticle is the main barrier limiting glyphosate activity (Perkins et al. 2008). Moreover, a large proportion of glyphosate isopropylamine is lost due to rain fastness, run-off/ erosion, volatilization, spray drift, and photodegradation. Finally, from the amount of pesticide applied, only less than 0.1% is estimated to reach the target sites (Hunsche et al. 2007; Wang and Liu 2007; Reichenberger et al. 2007). In order to minimize the pesticide dissipation and in order to increase the penetration of active ingredient into plant foliage, Lim et al. have described the development of environment friendly nanoemulsion system for the water-soluble herbicide glyphosate isopropylamine (Lim et al. 2012, 2013; Chaw Jiang et al. 2012; Jiang et al. 2011) Initially, the pre-formulation emulsion containing glyphosate isopropylamine concentrates was synthesized. Mixed surfactant systems were prepared by mixing

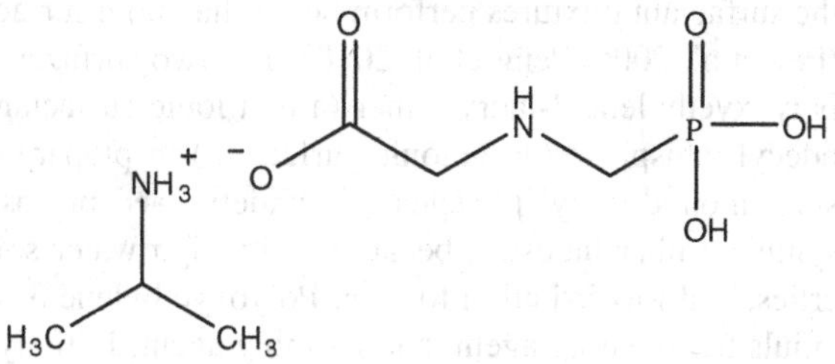

Fig. 11.9 Chemical structure of glyphosate isopropylamine

carbohydrate-derived alkylpolyglucoside and organosilicone surfactants. Two grades of alkylpolyglucoside surfactants were used, viz. short-chain alkylpoly-glucosides (SAPGs) with a mixture of alkyl-chain octyl and decyl at 45:55 (%w/w), respectively, and long-chain alkylpolyglucosides (LAPGs) with a mixture of alkyl-chain dodecyl, tetradecyl, and hexadecyl at 68, 26, and 6 (%w/w), respectively. The synthetic organosilicone surfactant wetting agents such as 3-(3-hydroxypropyl)-heptamethyltrisiloxane are known adjuvant for increasing the efficacy of glyphosate into plant tissue with greater weed control (Reddy and Singh 1992; Sharma and Singh 2000). Two types of esterified vegetable oils were used—short-chain fatty acid methyl esters (SFAMEs) containing a mixture of hexanoate, octanoate, decanoate, and laurate esters at 3.2, 50.8, 44.0, and 2.0 (%w/w), respectively, and long-chain fatty acid methyl esters (LFAMEs) containing a mixture of octanoate, decanoate, laurate, myristate, palmitate, stearate, oleate, linoleate, linolenate, and arachidate esters at 0.1, 0.1, 52.1, 17.7, 8.9, 2.2, 15.8, 2.8, 0.2, and 0.1 (%w/w), respectively. The oil phase of the short-chain and long-chain fatty acid methyl esters was further mixed by stirring with surfactant systems. Subsequently, the emulsions were formed by adding water to the mixtures. Finally, glyphosate isopropylamine was added to produce the pre-formulations. In the next phase, the nanoemulsion was developed from the pre-formulation concentrates by a low-energy emulsification method. The pre-formulation-concentrated samples were diluted with a large amount of water in the ratio of 1:200 (pre-formulation concentrate:water) along with stirring. The results showed characteristic difference in the flow ability (a physicochemical parameter of the pre-formulations). The pre-formulation prepared with short-chain alkylpolyglucoside surfactants exhibited Newtonian flow whereas with the increase in hydrophobic chain length of the surfactant (such as long-chain alkylpolyglucosides surfactant), a non-Newtonian or pseudo-plastic flow could be observed. The non-Newtonian system led to higher viscosity and flow resistance of the pre-formulations (showing greater resistance to sedimentation). The nanoemulsion formulations gave a low spray deposit as it exhibited a lower surface tension than the commercial herbicide formulation (Roundup®). The nanoemulsion system was expected to enhance the penetration for possible uptake of the herbicide. These nanoemulsions were further applied for successful control of certain weeds such as Indian goosegrass (*Eleusine indica*), creeping foxglove (*A. gangetica*), slender button weed (*D. ocimifolia*), and buffalo grass (*P. conjugatum*).

Liu et al. (2011) have conducted studies on the formation of bifenthrin oil-in-water nanoemulsions prepared with mixed surfactants. It has been observed that the surfactant mixtures perform better than pure surfactants to form nanoemulsions (Pey et al. 2006; Peng et al. 2010). The two surfactants selected in the study were polyoxyethylene 3-lauryl ether (a non-ionic surfactant) and di-potassium monod-odecyl phosphate (an anionic surfactant) to prepare the nanoemulsion. Di-potassium monododecyl phosphate is widely used in cosmetics industry, leather and synthetic fiber industry, because of its high water solubility, good foaming properties, and low irritation to skin. Polyoxyethylene 3-lauryl ether is widely used as emulsifier, wetting agent, and foaming agent. Initially, bifenthrin was dissolved in

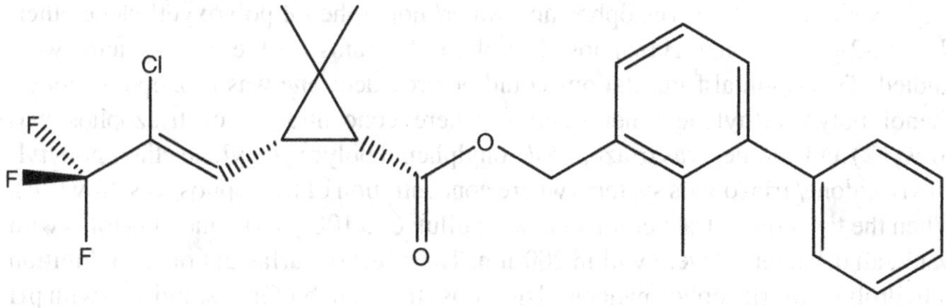

Fig. 11.10 Chemical structure of bifenthrin

dimethylbenzene, and then, the mixed surfactants were mixed together by stirring to form an oil phase. Nanoemulsions were prepared by adding the oil phase to water using a high-speed homogenizer operating at 3,000 rpm for 30 min at room temperature. The amount of bifenthrin (5 wt %) (Fig. 11.10) and the cosolvent dimethylbenzene (6 wt %) was kept constant and the amount of total surfactant was varied (8–12 wt %). The selection of appropriate surfactant HLB value was an important aspect for preparing stable emulsions. The mixing ratios of the two surfactants were adjusted to satisfy the proper HLB values for optimum emulsi-fication conditions (Peng et al. 2010). Using a single surfactant, di-potassium monododecyl phosphate or polyoxyethylene 3-lauryl ether, did not stabilize the O/W emulsion system (as indicated by their large mean droplet size). The addition of the second surfactant resulted in a subsequent decrease in the droplet size. It was reported that when polyoxyethylene 3-lauryl ether was used alone, the mean droplet size was found to be 465 nm and the size decreased to 136 nm by varying the surfactant ratio of di-potassium monododecyl phosphate: polyoxyethylene 3-lauryl ether to 1:9. Moreover, a gradual increase in mean droplet size was observed from 136 to 534 nm when the surfactant ratio was increased from 1:9 to pure di-potassium monododecyl phosphate alone. Finally, the most stable nanoemulsion was found with the di-potassium monododecyl phosphate: polyoxyethylene 3-lauryl ether ratio of 6:4. The optimized nanoemulsions showed high kinetic sta-bility without phase separation for 6 months at room temperature and the quality indicators of this nanoemulsion met the FAO standards.

Formulation and stability properties of a self-nanoemulsifying system were assessed for the delivery of triazophos pesticide (Fig. 11.11) (Song et al. 2009). The two systems selected for the pesticide entrapment were water/ nonylphenol

Fig. 11.11 Chemical structure of triazophos

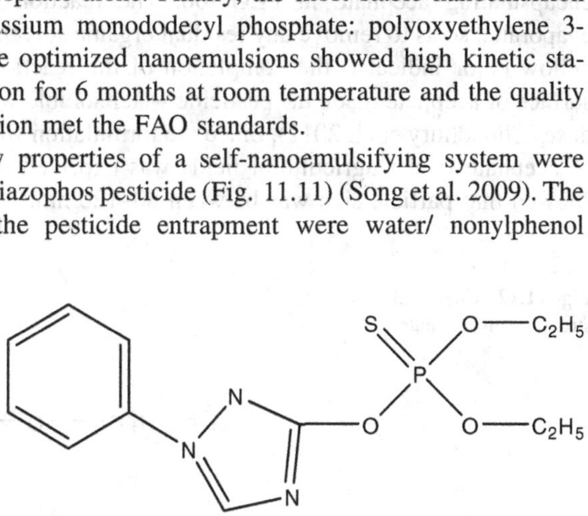

polyoxyethylene ether/ triazophos and water/ nonylphenol polyoxyethylene ether/ N-octyl-2-pyrrolidone/ triazophos. The phase diagrams for the both systems were studied. Two optimal formulations could be provided: one was triazophos/ nonyl-phenol polyoxyethylene ether system (where concentration of triazophos was 20 wt %) and another was triazophos/ nonylphenol polyoxyethylene ether/ N-octyl-2-pyrrolidone/ triazophos system (where concentration of triazophos was 25 wt %). When the two concentrate emulsions were diluted to 100-, 400-, and 800-folds with water, all the samples were within 200 nm. The effect of surfactant on the inhibition of hydrolysis of triazophos nanoemulsion was studied in buffered solutions (with pH 5, 7, and 9) and the results showed that triazophos was relatively stable in acidic and neutral solutions but got easily hydrolyzed in basic solutions.

11.1.4 Nanoencapsulation

Acephate (Fig. 11.12), a neurotoxic insecticide, is one of the most widely used organophosphorous insecticides in the Indian subcontinent. An attempt has been made to prepare an eco-friendly and hydrophilic formulation of acephate, as a core component, and a hydrophilic polymer polyethylene glycol-400 (PEG-400), as a surface stabilizer. Acephate contains a phosphate group and an amide linkage where the NH proton is capable of forming an H-bond with the carbonyl group and this has a major role in stabilizing the complex. Moreover, PEG-400 is a neutral ligand (with high HLB ratio) that can make the surface hydrophilic and induce a steric barrier by anchoring a long, mobile PEG chain on the surface of the core component, thereby exerting a protective action. A solution of PEG-400 in water was prepared by mixing them in the ratio of 9:1 under continuous stirring. To this, a 1 % (w/v) acephate solution in dichloromethane was poured under continuous stirring over a period of 4 h with heating at 45 °C. This procedure helped in encapsulating acephate in PEG-400. The reaction mixture was then vacuum evaporated so as to remove any residual organic solvent. The appearance of a pale-yellow color indicated the completion of the reaction. PEG-400 coating on the surface of acephate made the pesticide water soluble with potent activity at a lower dose (Choudhury et al. 2012). Since the formulation was water soluble, the toxicity of acephate in the agricultural fields was expected to get reduced. TEM image showed that particle size was between 80–120 nm.

Fig. 11.12 Chemical structure of acephate

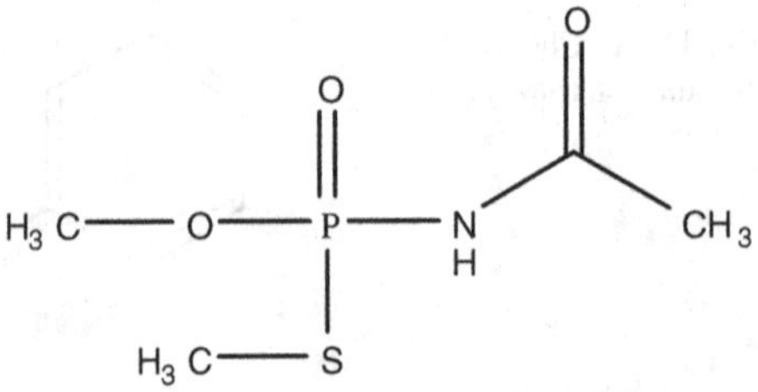

Furthermore, bioassay was carried out on two species of pests, *Spodoptera litura* and tea red spider mites (*Oligonychuscoffeae*) (Pradhan et al. 2013). Both the species were fed on castor leaves. The leaves were sprayed with fixed amounts of nanoencapsulated or commercial acephate of different concentration. Second instar larvae of *Spodoptera litura* were fed on the pesticide-treated castor leaves. Insecticide solutions used for nanoencapsulated acephate were 180, 240, and 300 ppm and compared with the commercial acephate preparation with the same concentration of acephate. The mortality was recorded after a week of spraying at the pupal stage and at the eclosion (adult emergence) stage. Control group of insects was treated with water only. The effectiveness of nanoacephate was also tested on tea red spider mites (*Oligonychuscoffeae*). Adult females were selected with a fine bristle brush onto castor leaves and maintained in plastic trays covered with nets. Mites were transferred from old leaves to new ones on all days of the experiment.

A high efficacy of nanoacephate against *S. litura* and mites (*Oligonychuscoffeae*) could be observed. The rate of mortality of pests treated with commercial acephate was not very significant as compared to the control at any stage of the experiments. In the case of *Spodoptera litura*, nearly 100 % mortality could be observed at 300 ppm, 75 % at 240 ppm, and 20 % at 180 ppm within a week.

This formulation was found to work in a controlled dose-dependent manner. Therefore, during adult emergence, not a single pest was left at 240 and 300 ppm. The higher concentration of nanoacephate also reduced fecundity of larvae when they reached adulthood. The same dose dependency was observed in the case of tea spider mites (*Oligonychuscoffeae*). In the case of tea spider mites, under *in vitro* conditions, nearly 100 % mortality could be observed at 300 ppm and 50 % at 180 ppm after 5 days of treatment. The field trials were also performed for control of *S. litura*, *Lipaphis erysimi* (mustard aphid) and *Bemisia tabaci* (whitefly) on castor leaves. The results for the developed nanoacephate formulations were compared with the results using commercial pesticide. Foliar spray of the nano-acephate at 180, 240, and 300 ppm gave good control of *S. litura* and *Lipaphis erysimi* as compared to the commercial one. Nanoformulation was moderately toxic to the whiteflies at single spray, as pests appeared after 5 days of treatment. The pests showed resistance to commercial acephate in all of the cases. The data obtained suggested that nanoacephate had excellent activity against a broad spectrum of agriculturally harmful pests both under *in vitro* and *in vivo* conditions.

Zhu et al. (2009) prepared a nanoencapsulation of β-Cypermethrin (Fig. 11.8) by complex coacervation in a microemulsion. The pesticide was dissolved and oil-in-water (O/W) microemulsion was used as a template for preparing nanocapsules by complex coacervation of biopolymers acacia and gelatin.

Two prerequisites must be satisfied so as to obtain a stable microemulsion. The oil phase must have a capability to solubilize β-cypermethrin, and the O/W microemulsion must remain stable in the temperature range of 4–40 °C and pH 3.5–9.0. Initially, the microemulsion was prepared by dissolving β-cypermethrin in butyl acetate, which acted as a solvent and also the oil phase for the micro-emulsion. Additionally, β-cypermethrin was much more soluble in butyl acetate

(69.1 g/100 mL) than in water (1.0 μg/100 mL). In-house synthesized alkyl polyglucoside (APG) (C10) was used as a surfactant, butanol as a cosolvent and the mixture composed with $WAPG_{:butanol} = 4:1$. The microemulsion, synthesized in an aqueous solution containing sodium salicylate, was found to be suitable for these conditions. The microemulsion was filtered before use and kept in a water-jacketed beaker for incubation at 40 °C. Next steps involved the coacervation reaction (Ichwan et al. 1999). Complex coacervation is a pH-dependent process, as the charge and charge density of polymers are expected to change with pH. Complex coacervation between acacia and gelatin is restricted to a narrow pH range because both molecules carry opposite charges. Acacia is always negatively charged in the solution while the charge of gelatin molecules depends on pH. Gelatin, being an amphoteric polymer, carries a positive (cationic) charge under acidic conditions and remains negatively (anionic) charged in alkaline pH. An isoelectric point is the pH at which the polymer carries an equal number of positively and negatively charged groups, according to the following balance:

The maximum coacervation is expected to correspond to the electrical equivalence pH, where both the polymers carry equal but opposite charges. At alkaline pH (9.0), the charges of both the gelatin solution and the acacia solution are negative, and as a result, there is no obvious coacervation between the two polymers. When the pH of the mixture is below 3.7, coacervation between gelatin and acacia solutions quickly results in the formation of a white deposit in the solution, claiming that the rate of coacervation is so rapid that this pH is not suitable for the nanocapsule preparation. At pH 4.8, coacervation between acacia and gelatin occurs in a controlled manner. Therefore, coacervation between gelatin and acacia was restricted to pH ranging between 3.7 and 4.8.

To the preformed microemulsion, first, an aqueous solution of gelatin was added followed by the addition of an aqueous solution of acacia. The pH of the solution was maintained at 3.8 by the addition of 0.1 M hydrochloric acid, and the temperature was adjusted to 4–6 °C by keeping it in an ice bath. The low temperature led to the hardening of the nanocapsules. Following this, the cross-linking agent (glutaraldehyde) solution was added slowly by stirring the suspension. The product was ultra-centrifuged and the supernatant liquid was discarded. The collected nanocapsules were then washed twice with double-distilled water, and it was centrifuged each time after washing. Being a pH-dependent process, the pH had an effect on the production of the nanocapsules. The yield of the nanocapsules was reported to be highest (around 40 %) at pH 3.8. Further increase in pH led to decrease in the yield.

At an electrical equivalence pH, the attractive forces between the charged components neutralized each other, leading to a strong binding and the highest coacervation yield. In this experiment, the pH value of 3.8 was near to the electrical equivalence pH, and so the highest yield of the nanocapsule was obtained. The entrapment efficiency of β-cypermethrin in the nanocapsule was also high (over 60 %). The encapsulation efficiency was not affected by increasing the amount of β-cypermethrin in the butyl acetate. Increasing the oil content and reducing the surfactant content of the microemulsion led to a decrease in the entrapment efficiency of the nanocapsule, and this may be because of the loose structure of the surfactant in the microemulsion droplet. The entrapment efficiency could be enhanced by increasing the concentration of gelatin and acacia in the microemulsion. This result may be explained by the fact that the wall of the nanocapsule was thickened by more addition of the polymers and a compact structure led to high-entrapment efficiency. The results showed that the nanocapsule prepared had a mean diameter below 100 nm with a good dispersion and had spherical morphology.

Nanocapsules of lansiumamide B (Fig. 11.13) were prepared by the microemulsion polymerization method so as to increase its nematicidal efficacy (Yin et al. 2012). The process included sodium dodecyl sulfate as an emulsifier and N-amyl alcohol (as an auxiliary emulsifier) which were dissolved in water. Lansiumamide B was dissolved in a small amount of petroleum ether and chloroform in the precursor solution, into which methyl methacrylate and styrene monomers were added. The solution was ultrasonicated to obtain a clarified emulsion which was then transferred into a three-necked flask. Finally, the nanoencapsulation of the preformed emulsion is done by using azobisisobutyronitrile (AIBN) as an initiator. The azobisisobutyronitrile induced the production of free radical leading to the copolymerization of methyl methacrylate and styrene. The average diameter of the particles was around 38 nm with a narrow particle size distribution. The encapsulation efficiency and loading efficiency of Lansiumamide B in the nanocapsules were reported to be 96 and 49 %, respectively.

The efficacy of the lansiumamide B loaded nanocapsules was evaluated by estimating the nematicidal activity on *Bursaphelenehus xylophilus* (pinewood

Fig. 11.13 Chemical structure of lansiumamide B (Matsui et al. 2013)

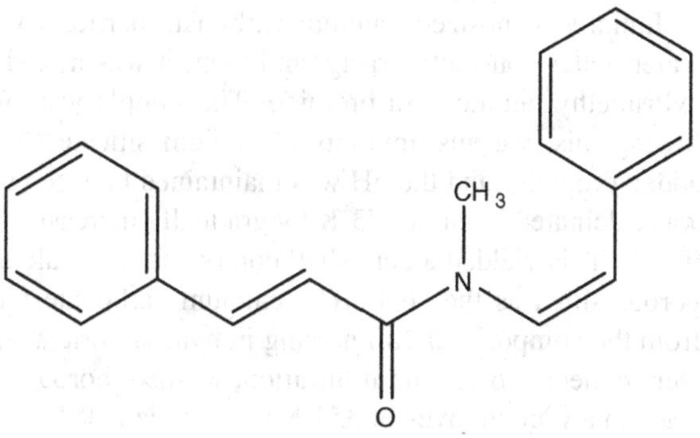

nematode) and *Meloidogyne incognita* (Southern root-knot nematode). *B. xylophilus* is the causal agent of the pine wilt disease, which causes severe ecological and economic losses in coniferous forests (Pereira et al. 2013). *M. incognita* in the second-stage juveniles (J2) can penetrate the root and infect the plant (Abad et al. 2008). Mortality of nematodes was counted only when their bodies were straight and they did not move when poked (Oka et al. 2009). The LC50 for *B. xylophilus* was found to be 4.9 and 2.1 mg/L for plain and nanoencapsulated lansiumamide B, respectively. In the case of J2 of *M. incognita*, the LC50 was found to be 24.4 and 19.4 mg/L for plain and nanoencapsulated lansiumamide B, respectively. It could be clearly seen that in a 24-h run, the dose requirement was 2.29-fold less for *B. xylophilus* and 1.26-fold less for J2 of *M. incognita* for the nanoformulation as compared to the plain lansiumamide B. In another experiment, pre-sterilized soil was used to cultivate *Pomoea aquatica* planted in a pot. Each pot was infected by J2 of *M. incognita* suspension by digging holes. The results suggested that nanoencapsulated lansiumamide B formulation performed more efficiently and provides longer effective maintenance against plant parasitic nematodes.

11.1.5 Nanosilica

Many pesticides are sensitive to UV-light and their half-life time is very short, such as avermectin (6 h) and phoxim (40 min). The list also includes isoprothiolane, dimethachlon, bentazone. It is, therefore, necessary to encapsulate the active ingredients into some form of sunscreen carriers for the protection of the labile compounds against photo-degradation and in the process also slow down the release of the entrapped compounds. The preparation of porous hollow silica nanoparticles (PHSNs) with various shell thicknesses in the range of 5–45 nm, and a pore diameter of about 4–5 nm is being been described for the same purpose. PHSNs have been synthesized by a sol-gel route with two different structure-directing templates, and their shell thickness has been controlled by adjusting the reactant ratio of sodium silicate/calcium carbonate. PHSNs can protect the model pesticide avermectin against photo-degradation effectively (Li et al. 2006). The UV-shielding property can be further improved by increasing the shell thickness.

Initially, nanosized calcium carbonate particles were suspended into distilled water under constant stirring, and then, it was mixed with the surfactant hexadecyltrimethyl ammonium bromide. The sample was further heated to 353 K. Following this, various amounts of sodium silicate ($Na_2SiO_3.9H_2O$) solution were added dropwise and the pH was maintained between 9 and 10. Finally, the powder was calcinated in air at 973 K by gradually increasing the temperature from 298 to 973 K. This yielded a core shell composite with calcium carbonate as the core and porous silica as the shell. The calcium carbonate templates were then removed from the composite by immersing in hydrochloric acid solution. The nanoparticles were collected by vacuum filtration, washed thoroughly with deionized water, and dried in a vacuum oven at 353 K to produce PHSNs. The pesticide avermectin was loaded by using supercritical fluid technology. Carbon dioxide was selected for non-polar supercritical fluid and acetone was added in the adsorption experiments as a cosolvent for the loading of avermectin on PHSNs (due to avermectin's high solubility in acetone). The mixture was put into a high-pressure adsorption apparatus at 308 K and 15 MPa for 6 h. Finally, the resulting powder was dried in a vacuum oven to obtain the avermectin-loaded PHSNs (Li et al. 2006, 2007) The PHSN carriers could load 60% w/w of avermectin and the amount of loaded avermectin (Fig. 11.14) decreased with the increasing of shell thickness. Lie et al used similar process to develop PHSN for pesticide delivery.

The UV-shielding properties of the produced PHSNs have also been analyzed. The avermectin-loaded PHSNs were added to mixture of ethanol/water (30:70, v/v) and this mixture was then transferred into a chamber that contained a 125 W (Emax = 365 nm) UV-lamp as a UV-light source. A constant temperature (298 K) was maintained throughout the experiment. It was found that in control experiments, un-encapsulated avermectin was completely decomposed after UV-irradiation for 2 h, whereas the pesticide was still detectable in the solution of avermectin-PHSNs even after 12 h, indicating that the PHSN carriers could protect avermectin from UV-degradation. As the thickness of the shell was increased, the release of avermectin became much slower. Decreasing the speed of release of the remaining avermectin into the solution provided a better protection, indicating that the UV-shielding properties of PHSN carriers for avermectin could be improved by increasing the shell thickness.

The study of controlled release of loaded avermectin pesticide from PHSN samples has also been investigated. The release medium was an ethanol/water mixture (30:70 v/v) at pH 7.0 and the stirring speed was 100 rpm. The release experiment also confirmed slower avermectin release after an initial burst. The initial burst release was caused by the dissolution of the avermectin loaded on the external surface of the PHSNs and was independent of the shell thickness. Thicker

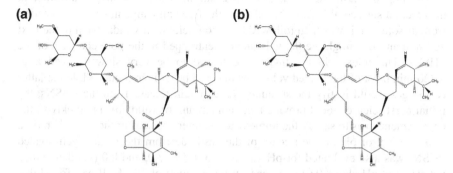

Fig. 11.14 Molecular structure of avermectin **a** Avermectin B1$_a$ and **b** Avermectin B1$_b$ (Putter et al. 1981)

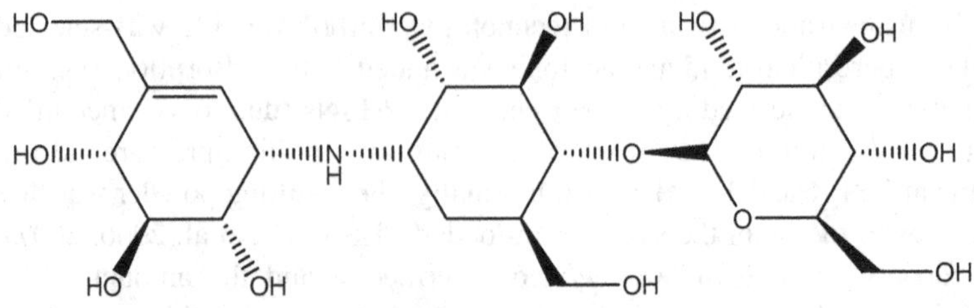

Fig. 11.15 Chemical structure of validamycin (Liu et al. 2006)

shell resulted in an enhanced resistance to the avermectin diffusion across the pore channels, leading to a slower avermectin release. Therefore, the release rate of avermectin from avermectin-loaded PHSNs could be controlled by adjusting the shell thickness (Li et al. 2006).

A similar sol-gel method has also been used to develop validamycin (Fig. 11.15)-entrapped PHSNs. Liu el al. demonstrated that nanosized hollow silica particles with nearly uniform shell thickness could be used for this purpose (Liu et al. 2006). It was estimated that the interior size of the PHSNs was about 80 nm and the shell thickness was about 15 nm. The loading capacity of 25 % could be achieved for validamycin by the simple immersing method for 14 days, whereas loading capacity of 36 % could be achieved by the supercritical fluid-loading technique in about 9 h. PHSNs had a porous hollow structure and the high pressure exerted by the supercritical fluids (without using any aqueous solution) forced the pesticide to penetrate and get entrapped by entering the inner core of PHSNs. During the dissolution test, it was found that free validamycin (adsorbed on the surface of the PHSNs) dissolved immediately after they were added into the dissolution apparatus. On the other hand, the validamycin loaded inside the PHSNs remained entrapped during the experiment. In the case of validamycin-loaded PHSNs, 65 % of the loaded validamycin could be released into the dissolution medium immediately after the validamycin-loaded PHSNs were added and 10 % of the loaded validamycin could be released in the next 10 min. This release of 75 % validamycin might be due to the dissolution of the validamycin adsorbed at the external surface of PHSNs, which, like the free validamycin, had no protection from dissolution. It was then followed by a slow release of validamycin in the next 700 min and this might be the validamycin entrapped in the pore channels of the PHSNs. The release in the final stage turned to be very slow as even after 4,320 min most of the loaded validamycin could be delivered out. While the latter two stages would justify the sustained validamycin release from the PHSNs, the initial burst release would provide a certain amount of validamycin quickly to the environment so as to satisfy the immediate treatment need after the administration.

The effect of pH on the release of the pesticide from the validamycin-loaded PHSNs was also evaluated for pH values of 3.0, 5.0, 7.0, and 8.0 (validamycin is not stable at pH above 9.0) at a constant temperature of 298 K. It was found that after 700 min, 78 and 91 % of validamycin was released at the pH of 3.0 and 8.0, respectively. The molybdenum blue test confirmed that the PHSNs could be partially dissolved in basic solutions facilitating the validamycin release. The multi-stage release behavior makes the validamycin-loaded PHSNs a promising carrier in agriculture, especially for pesticides whose immediate as well as prolonged release is needed.

References

Abad P, Gouzy J, Aury J-M, Castagnone-Sereno P, Danchin EG, Deleury E, Perfus-Barbeoch L, Anthouard V, Artiguenave F, Blok VC (2008) Genome sequence of the metazoan plant-parasitic nematode Meloidogyne incognita. Nat Biotechnol 26(8):909–915

Ahmad SN, Islam MT, Abdullah DK, Omar D (2012) Evaluation of physicochemical characteristics of microemulsion formulation of rotenone and its insecticidal efficacy against Plutella xylostella L.(Lepidoptera: Plutellidae). Food, Agriculture and Environment (JFAE) 10(3&4):384–388

Amrhein N, Schab J, Steinrücken H (1980) The mode of action of the herbicide glyphosate. Naturwissenschaften 67(7):356–357

Anees AM (2008) Larvicidal activity of Ocimum sanctum Linn. (Labiatae) against Aedes aegypti (L.) and Culex quinquefasciatus (Say). Parasitol Res 103(6):1451–1453

Anjali CH, Sudheer Khan S, Margulis-Goshen K, Magdassi S, Mukherjee A, Chandrasekaran N (2010) Formulation of water-dispersible nanopermethrin for larvicidal applications. Ecotoxicol Environ Saf 73(8):1932–1936

Bentley MD, Hassanali A, Lwande W, Njoroge PEW, Sitayo ENO, Yatagai M (1987) Insect antifeedants from Tephrosia elata Deflers. Int J Trop Insect Sci 8(01):85–88. doi:10.1017/S1742758400007025

Chaw Jiang L, Basri M, Omar D, Abdul Rahman MB, Salleh AB, Raja Abdul Rahman RNZ, Selamat A (2012) Green nano-emulsion intervention for water-soluble glyphosate isopropylamine (IPA) formulations in controlling Eleusine indica (E. indica). Pestic Biochem Physiol 102(1):19–29. doi:10.1016/j.pestbp.2011.10.004

Chhetri AB, Tango MS, Budge SM, Watts KC, Islam MR (2008) Non-edible plant oils as new sources for biodiesel production. Int J Mol Sci 9(2):169–180

Choudhury SR, Pradhan S, Goswami A (2012) Preparation and characterisation of acephate nano-encapsulated complex. Nanosci Meth 1(1):9–15

Cognis G (2007) Agnique PG www.cognis.com. Accessed 2 May 2013

Crombie L (1999) Natural product chemistry and its part in the defence against insects and fungi in agriculture. Pestic Sci 55(8):761–774. doi:10.1002/(sici)1096-9063(199908)55:8<761:aid-ps26>3.0.co;2-2

European-Pharmacopoeia (2005) European Pharmacopoeia vol 5.0. Convention on the Elaboration of a European Pharmacopoeia. Council of Europe

EXTOXNET (1995) Pesticide Information Profile: Azadirachtin. http://pmep.cce.cornell.edu/profiles/extoxnet/24d-captan/azadirachtin-ext.html#4. Accessed 2 May 2013

Fradin MS, Day JF (2002) Comparative efficacy of insect repellents against mosquito bites. N Engl J Med 347(1):13–18. doi:10.1056/NEJMoa011699

Guadaño A, Gutiérrez C, de la Peña E, Cortes D, González-Coloma A (2000) Insecticidal and mutagenic evaluation of two Annonaceous Acetogenins. J Nat Prod 63(6):773–776. doi:10.1021/np990328+

Hemsley AR, Poole I (2004) The evolution of plant physiology, vol 21. Elsevier

Hu M, Zhong G, Sun ZT, Sh G, Liu H, Liu X (2005) Insecticidal activities of secondary metabolites of endophytic Pencillium sp. in Derris elliptica Benth. J Appl Entomol 129(8):413–417

Hunsche M, Damerow L, Schmitz-Eiberger M, Noga G (2007) Mancozeb wash-off from apple seedlings by simulated rainfall as affected by drying time of fungicide deposit and rain characteristics. Crop Prot 26(5):768–774

Ichwan AM, Karimi M, Dash AK (1999) Use of gelatin–acacia coacervate containing benzocaine in topical formulations. J Pharm Sci 88(8):763–766

Jiang LC, Basri M, Omar D, Rahman MBA, Salleh AB, Rahman RNZRA (2011) physicochemical characterization of nonionic surfactants in oil-in-water (O/W) nano-emulsions for new pesticide formulations. Int J Appl Sci Technol 1(5)

Koundal K, Rajendran P (2003) Plant insecticidal proteins and their potential for developing transgenics resistant to insect pests. Indian J Biotechnol 2(1):110–120

Lawrence MJ, Rees GD (2000) Microemulsion-based media as novel drug delivery systems. Adv Drug Deliv Rev 45(1):89–121. doi:10.1016/S0169-409X(00)00103-4

Lee JB, Lee SH (2011) Dynamic wetting and spreading characteristics of a liquid droplet impinging on hydrophobic textured surfaces. Langmuir 27(11):6565–6573. doi:10.1021/la104829x

Li H, Geng S (2013) Development and characterization of microsatellite markers for Derris elliptica (Fabaceae), an insecticide-producing plant. Sci Hortic 154:54–60. doi:10.1016/j.scienta.2013.02.026

Li Z-Z, Xu S-A, Wen L-X, Liu F, Liu A-Q, Wang Q, Sun H-Y, Yu W, Chen J-F (2006) Controlled release of avermectin from porous hollow silica nanoparticles: Influence of shell thickness on loading efficiency, UV-shielding property and release. J Controlled Release 111(1–2):81–88. doi:10.1016/j.jconrel.2005.10.020

Li Z-Z, Chen J-F, Liu F, Liu A-Q, Wang Q, Sun H-Y, Wen L-X (2007) Study of UV-shielding properties of novel porous hollow silica nanoparticle carriers for avermectin. Pest Manag Sci 63(3):241–246. doi:10.1002/ps.1301

Lim CJ, Basri M, Omar D, Abdul Rahman MB, Salleh AB, Raja Abdul Rahman RNZ (2012) Physicochemical characterization and formation of glyphosate-laden nano-emulsion for herbicide formulation. Ind Crops Prod 36(1):607–613. doi:10.1016/j.indcrop.2011.11.005

Lim CJ, Basri M, Omar D, Abdul Rahman MB, Salleh AB, Raja Abdul Rahman RNZ (2013) Green nanoemulsion-laden glyphosate isopropylamine formulation in suppressing creeping foxglove (A. gangetica), slender button weed (D. ocimifolia) and buffalo grass (P. conjugatum). Pest Manage Sci 69(1):104–111. doi:10.1002/ps.3371

Liu H, Cupp EW, Micher KM, Guo A, Liu N (2004) Insecticide resistance and cross-resistance in Alabama and Florida strains of Culex quinquefaciatus. J Med Entomol 41(3):408–413

Liu F, Wen L-X, Li Z-Z, Yu W, Sun H-Y, Chen J-F (2006) Porous hollow silica nanoparticles as controlled delivery system for water-soluble pesticide. Mater Res Bull 41(12):2268–2275

Liu Y, Wei F, Wang Y, Zhu G (2011) Studies on the formation of bifenthrin oil-in-water nano-emulsions prepared with mixed surfactants. Colloids Surf, A 389(1–3):90–96. doi:10.1016/j.colsurfa.2011.08.045

Maia MF, Moore SJ (2011) Plant-based insect repellents: a review of their efficacy, development and testing. Malar J 10(Suppl 1):S11

Malmsten M (2002) Surfactants and polymers in drug delivery, vol 122. Drugs and the pharmaceutical sciences. Marcel Dekker, Inc, New York

Masmoudi H, Piccerelle P, Le Dréau Y, Kister J (2006) A rheological method to evaluate the physical stability of highly viscous pharmaceutical oil-in-water emulsions. Pharm Res 23(8):1937–1947

Matsui T, Ito C, Furukawa H, Okada T, Itoigawa M (2013) Lansiumamide B and SB-204900 isolated from Clausena lansium inhibit histamine and TNF-α release from RBL-2H3 cells. Inflamm Res 62(3):333–341. doi:10.1007/s00011-012-0586-8

Metcalf RL (1948) The mode of action of organic insecticides, vol 1–5. National Academies

Mohibbe Azam M, Waris A, Nahar NM (2005) Prospects and potential of fatty acid methyl esters of some non-traditional seed oils for use as biodiesel in India. Biomass Bioenergy 29(4):293–302. doi:10.1016/j.biombioe.2005.05.001

Nawrot J, Harmatha J, Kostova I, Ognyanov I (1989) Antifeeding activity of rotenone and some derivatives towards selected insect storage pests. Biochem Syst Ecol 17(1):55–57. doi:10.1016/0305-1978(89)90043-4

Nerio LS, Olivero-Verbel J, Stashenko E (2010) Repellent activity of essential oils: a review. Bioresour Technol 101(1):372–378. doi:10.1016/j.biortech.2009.07.048

Nisbet AJ (2000) Azadirachtin from the neem tree Azadirachta indica: its action against insects. Anais da Sociedade Entomológica do Brasil 29(4):615–632

Oka Y, Shuker S, Tkachi N (2009) Nematicidal efficacy of MCW-2, a new nematicide of the fluoroalkenyl group, against the root-knot nematode Meloidogyne javanica. Pest Manag Sci 65(10):1082–1089. doi:10.1002/ps.1796

Patravale V, Kulkarni R (2004) Nanosuspensions: a promising drug delivery strategy. J Pharm Pharmacol 56(7):827–840

Peng L-C, Liu C-H, Kwan C-C, Huang K-F (2010) Optimization of water-in-oil nanoemulsions by mixed surfactants. Colloids Surf, A 370(1–3):136–142. doi:10.1016/j.colsurfa.2010.08.060

Pereira F, Moreira C, Fonseca L, van Asch B, Mota M, Abrantes I, Amorim A (2013) New insights into the phylogeny and worldwide dispersion of two closely related nematode species, Bursaphelenchus xylophilus and Bursaphelenchus mucronatus. PLoS ONE 8(2):e56288. doi:10.1371/journal.pone.0056288

Perkins M, Bell G, Briggs D, Davies M, Friedman A, Hart C, Roberts C, Rutten F (2008) The application of ToF-SIMS to the analysis of herbicide formulation penetration into and through leaf cuticles. Colloids Surf, B 67(1):1–13

Pey CM, Maestro A, Solé I, González C, Solans C, Gutiérrez JM (2006) Optimization of nano-emulsions prepared by low-energy emulsification methods at constant temperature using a factorial design study. Colloids Surf, A 288(1–3):144–150. doi:10.1016/j.colsurfa.2006.02.026

Pradhan S, Roy I, Lodh G, Patra P, Choudhury SR, Samanta A, Goswami A (2013) Entomotoxicity and biosafety assessment of PEGylated acephate nanoparticles: A biologically safe alternative to neurotoxic pesticides. J Environ Sci Health, Part B 48(7):559–569

Pratap AP, Bhowmick D (2008) Pesticides as microemulsion formulations. J Dispersion Sci Technol 29(9):1325–1330

Prince LM (1975) Microemulsions versus micelles. J Colloid Interface Sci 52(1):182–188. doi:10.1016/0021-9797(75)90315-X

Putter I, Mac Connell J, Preiser F, Haidri A, Ristich S, Dybas R (1981) Avermectins: novel insecticides, acaricides and nematicides from a soil microorganism. Experientia 37(9):963–964

Raizada RB, Srivastava MK, Kaushal RA, Singh RP (2001) Azadirachtin, a neem biopesticide: subchronic toxicity assessment in rats. Food Chem Toxicol 39(5):477–483. doi:10.1016/S0278-6915(00)00153-8

Rane SS, Anderson BD (2008) What determines drug solubility in lipid vehicles: is it predictable? Adv Drug Deliv Rev 60(6):638–656. doi:10.1016/j.addr.2007.10.015

Reddy KN, Singh M (1992) Organosilicone adjuvants increased the efficacy of glyphosate for control of weeds in citrus (Citrus spp.). HortScience 27(9):1003–1005

Reichenberger S, Bach M, Skitschak A, Frede H-G (2007) Mitigation strategies to reduce pesticide inputs into ground-and surface water and their effectiveness: a review. Sci Total Environ 384(1):1–35

Sakulku U, Nuchuchua O, Uawongyart N, Puttipipatkhachorn S, Soottitantawat A, Ruktanonchai U (2009) Characterization and mosquito repellent activity of citronella oil nanoemulsion. Int J Pharm 372(1–2):105–111. doi:10.1016/j.ijpharm.2008.12.029

Schulman JH, Stoeckenius W, Prince LM (1959) Mechanism of formation and structure of micro emulsions by electron microscopy. J Phys Chem 63(10):1677–1680. doi:10.1021/j150580a027

Sharma S, Singh M (2000) Optimizing foliar activity of glyphosate on Bidens frondosa and Panicum maximum with different adjuvant types. Weed Res 40(6):523–533

Singla M, Patanjali PK (2013) Phase behaviour of neem oil based microemulsion formulations. Ind Crops Prod 44:421–426. doi:10.1016/j.indcrop.2012.10.016

Song S, Liu X, Jiang J, Qian Y, Zhang N, Wu Q (2009) Stability of triazophos in self-nanoemulsifying pesticide delivery system. Colloids Surf, A 350(1–3):57–62. doi:10.1016/j.colsurfa.2009.08.034

Steinrücken H, Amrhein N (1980) The herbicide glyphosate is a potent inhibitor of 5-enolpyruvylshikimic acid-3-phosphate synthase. Biochem Biophys Res Commun 94(4):1207–1212

Talegaonkar S, Azeem A, Ahmad FJ, Khar RK, Pathan SA, Khan ZI (2008) Microemulsions: a novel approach to enhanced drug delivery. Recent Pat Drug Delivery Formulation 2(3):238–257

Tamhane VA, Chougule NP, Giri AP, Dixit AR, Sainani MN, Gupta VS (2005) In vivo and in vitro effect of Capsicum annum proteinase inhibitors on Helicoverpa armigera gut proteinases. Biochimica et Biophysica Acta (BBA)—General Subjects 1722(2):156–167. doi:10.1016/j.bbagen.2004.12.017

Tamhane VA, Dhaware DG, Khandelwal N, Giri AP, Panchagnula V (2012) Enhanced permeation, leaf retention and plant protease inhibitor activity with bicontinuous microemulsions. J Colloid Interface Sci

Tomlin CDS (2000) The pesticide manual. BCPC, Farnham

Trongtokit Y, Rongsriyam Y, Komalamisra N, Apiwathnasorn C (2005) Comparative repellency of 38 essential oils against mosquito bites. Phytotherapy Res 19(4):303–309. doi:10.1002/ptr.1637

Tuffi Santos LD, Meira RMSA, Ferreira FA, Sant'Anna-Santos BF, Ferreira LR (2007) Morphological responses of different eucalypt clones submitted to glyphosate drift. Environ Exp Bot 59(1):11–20

USEPA-OPP (2009) Cold Pressed Neem Oil PC Code 025006. U.S. Environmental Protection Agency Office of Pesticide Programs. http://www.epa.gov/opp00001/chem_search/reg_actions/registration/decision_PC-025006_14-Oct-09.pdf. Accessed 5 June 2013

Vadillo DC, Soucemarianadin A, Delattre C, Roux DCD (2009) Dynamic contact angle effects onto the maximum drop impact spreading on solid surfaces. Phys Fluids 21(12):122002–122008

Wang C, Liu Z (2007) Foliar uptake of pesticides—present status and future challenge. Pestic Biochem Physiol 87(1):1–8

Wang L, Li X, Zhang G, Dong J, Eastoe J (2007) Oil-in-water nanoemulsions for pesticide formulations. J Colloid Interface Sci 314(1):230–235

Wheeler GS, Slansky F, Yu SJ (2001) Food consumption, utilization and detoxification enzyme activity of larvae of three polyphagous noctuid moth species when fed the botanical insecticide rotenone. Entomol Exp Appl 98(2):225–239. doi:10.1046/j.1570-7458.2001.00778.x

Yin Y-H, Guo Q-M, Han Y, Wang L-J, Wan S-Q (2012) Preparation, characterization and nematicidal activity of lansiumamide B nano-capsules. J Integr Agric 11(7):1151–1158

Zhang L, Han J-J, Li J-J, Liu T-Q (2013) Properties and spreading kinetics of water-based cypermethrin microemulsions. Acta Phys Chim Sin 29(2):346–350

Zhu Y, An X, Li S, Yu S (2009) Nanoencapsulation of β-cypermethrin by complex coacervation in a microemulsion. J Surfactants Deterg 12(4):305–311

Chapter 12
Characterization and In Vitro Release Techniques for Nanoparticulate Systems

Following a successful synthesis of nanoparticle, it becomes necessary to define the size and quality of the nanoproducts formed. Modern analytical techniques can handle this responsibility with greatest accuracy, and a few of them are being discussed briefly.

12.1 Dynamic Light Scattering

Dynamic light scattering (DLS) is also known as photon correlation spectroscopy (PCS) and quasi-elastic light-scattering (QELS) technique (NBTC; PHOTOCOR; LSinstruments). DLS is one of the most popular light-scattering techniques used today because it is a rapid technique that allows detection of particle size up to 1 nm in diameter. DLS technique can be used for particle size determination of emulsions, micelles, polymers, proteins, nanoparticles, or colloids (LSinstruments). DLS works on a very simple and basic principle that when the sample is illuminated by a laser beam, the particle can scatter the light in all directions, and the fluctuations of the scattered light is detected at a known scattering angle θ by a fast photon detector.

The instrumentation analysis of DLS is based on the following assumption:

- The dispersed or suspended particles in a liquid medium undergo Brownian motion, which can cause fluctuations of local concentration of the particles, resulting in local inhomogeneities of the refractive index. This in turn can result in the fluctuations of intensity of the scattered light.
- The diffusion coefficient of the particles is inversely proportional to the decay time of light-scattering fluctuations. The decay time can be obtained from the time-dependent correlation function of the scattered light.
- The particle size can be calculated in accordance with the Stokes–Einstein formula relating the particle size to the diffusion coefficient and viscosity.

The DLS method consists of determining the velocity distribution of the particle movement by measuring dynamic fluctuations of the intensity of the scattered

A. De et al., *Targeted Delivery of Pesticides Using Biodegradable Polymeric Nanoparticles*, SpringerBriefs in Molecular Science, DOI: 10.1007/978-81-322-1689-6_12, © The Author(s) 2014

light. The dispersed particles or macromolecules suspended in a liquid medium can undergo Brownian motion, causing the fluctuations in the local concentration of the particles and resulting in local inhomogeneities of the refractive index. This in turn can result in the fluctuations of the intensity of the scattered light. The linewidth of the light-scattered spectrum Γ (defined as the half width at half maximum) is proportional to the diffusion coefficient of the particles D (Eq. 12.1):

$$\Gamma = D_t q^2 \qquad (12.1)$$

where

$$q = \frac{4\pi n}{\lambda} \sin\left(\frac{\theta}{2}\right) \qquad (12.2)$$

n is the refractive index of the medium, λ the laser wavelength, and θ the scattering angle. Using the assumption that the particles are spherical and non-interacting, the mean radius can be obtained from the Stokes–Einstein equation (Eq. 12.3).

$$R_h = \frac{k_B T}{6\pi\eta D} \qquad (12.3)$$

where k_B is the Boltzmann constant, T the absolute temperature, and η the shear viscosity of the solvent. Information about the light-scattering spectrum can be obtained from the autocorrelation function $G(\tau)$ of the light-scattering intensity. In the simplest case of spherical, monodisperse, and non-interacting particles in a dust-free fluid, the characteristic decay time of the correlation function is inversely proportional to the linewidth of the spectrum. Therefore, the diffusion coefficient and the particle size or viscosity can be found by fitting the measured correlation function to a single exponential function.

There exist two techniques of measuring the correlation function: heterodyning and homodyning. In heterodyne measurements (which are most suitable for small intensities), the scattered light is mixed coherently with a static light source at the incident wavelength and the static field is added to the scattered fields at the photodetector. Equation 12.1 that connects the linewidth Γ and the diffusion coefficient D is given for a heterodyne spectrum. In homodyne measurements, the photodetector receives scattered light only. Homodyning is most suitable for strong intensities (e.g., near the critical point of the fluid or for colloid systems). In the case of a homodyne spectrum, the connection between Γ and D reads

$$\Gamma = 2D_t q^2 \qquad (12.4)$$

The DLS instruments that measure at a fixed angle can determine the mean particle size in a limited size range. More elaborate multiangle instruments can determine the full particle size distribution.

A DLS measurement for estimation of silica nanoparticles for the average mean size and size distribution of the powder can be performed using Photocor-FC. The powder was dispersed in water and sonicated for 5 min and then kept in argon ion

Fig. 12.1 DLS picture of silica nanoparticle suspended in distilled water

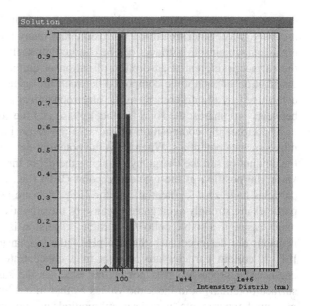

laser operating at 633 nm as a light source. Detection of diffracted light was done at 90° to an incident laser beam. Diffractogram was obtained with the help of Photocor and Dynals software. Figure 12.1 shows the size of nanosilica at around 105 nm. The suspended nanoparticles were found to be highly monodispersed as the polydispersibility index (PDI) is found to be 0.419.

12.2 Zeta Potential

A charged colloidal particle suspended in an electrolyte solution is surrounded by a cloud of counterions. The set of surface charges and countercharges is called the electrical double layer (Contescu 2009). The electrical double layer plays an essential role in various interfacial electrical phenomena on the particle surface and in the particle–particle interaction in a colloid suspension. Generally, it is almost impossible to measure the surface potential on colloid particles. However, we can measure the potential near the particle surface. It is called the zeta (ζ) potential. The zeta potential is the potential at the hydrodynamic slipping plane in the electrical double layer; hence, its value is not precisely the same as that of making a stable suspension because the total interaction potential between two particles (a bit distant from their surface) is essential for a stable dispersion. The ζ-potential has been considered to provide useful information necessary for preparing stable colloidal suspensions in many application fields including food preparation, agriculture, pharmaceuticals, paper industry, ceramics, paints, coatings, photographic emulsions, etc. The concept of the zeta potential is also very important in such diverse processes as environmental transport of nutrients, sol–gel synthesis, mineral recovery, wastewater treatment, corrosion, and many more. There are several origins from which solid surfaces can be charged: dissociation of chemical groups on the surface, preferential adsorption of cation or anion onto the surface, etc. The distribution of each ionic compound between the surface and the solution bulk is determined by the differences in the electrochemical potential of each compound between two phases: the solid (surface) phase and the solution phase.

Therefore, the composition of the solution is an important factor that determines surface potentials. When H^+ is the potential-determining ion, we can change the amount of surface charge by changing the pH of the solution. It is important to know the position of the isoelectric point (IEP) (i.e., the pH value at which the particles have zero ζ-potential). At the IEP, there are no repulsive forces and the particles are strongly aggregated because of the attractive van der Waals forces. In many cases, if a stable colloidal particle dispersion is desired, the colloidal suspensions are designed such that the pH of the suspension is well away from the IEP. If colloid particles are brought to a concentrated situation through some engineering processes, it is not certain whether the surface charges (and hence surface potential) hold the same values as that in a diluted dispersion. It must be measured experimentally, and several methods have been explored in recent years.

Nanoparticles with a zeta potential between -10 mV and $+10$ mV are considered approximately neutral, while nanoparticles with zeta potentials greater than $+30$ mV or less than -30 mV are considered strongly cationic and strongly anionic, respectively. Since most cellular membranes are negatively charged, zeta potential can affect a nanoparticle's tendency to permeate membranes, with cationic particles generally displaying more toxicity associated with cell wall disruption (McNeil 2010).

12.3 Atomic Force Microscope

The invention of the atomic force microscope (AFM) which is also known as scanning probe microscope (SPM) has revolutionized the scientific study of surfaces. Metallic or polymeric thin-film surfaces can be now imaged with nanometer resolution. Moreover, nowadays, it is possible to image biological materials such as DNA, proteins, and bacteria in their own natural environment.

The basic machineries of an AFM are as follows:

- Helium–neon laser
- An AFM probe
- A piezoelectric scanner
- A photodiode detector

A laser light is reflected off the top of a probe and is detected by a photodiode detector. The probe consists of a sharp tip attached to a flexible cantilever beam. The piezoelectric scanner controls either the vertical position of the surface or the vertical position of the probe. The piezoelectric scanner is able to move in three dimensions with angstrom-level precision. When the surface and the tip are brought closer together, the interaction between them causes the tip to be deflected either toward the surface (because of attractive forces) or away from the surface (because of repulsive forces). This deflection of the tip is recorded as the change in the position of the laser on the detector. The piezo responds to this change by increasing or decreasing the height of the sample to maintain a constant distance between the tip and the surface. Images can be recorded as either the deflection on the detector (deflection image) or the vertical distance of the piezo (height image). The vertical resolution of AFM images is dictated by the interaction between the tip and the surface. The lateral resolution is determined by the size of the tip. In most systems, the cantilever is tilted approximately 12° toward the surface. At a 0° scan angle, the tip moves across the surface. The AFM tips have traditionally been pyramidal in shape, but can also be of other shapes, such as a nanowire or a colloidal sphere.

The image produced by the AFM is a convolution of both the shape of the surface being imaged and the tip geometry. The rule of thumb is that the image will accurately reflect the surface structure if the difference in the image of the radius of curvature of the tip is one-tenth the radius of the imaged structure. The exception to this rule occurs if the height of the imaged structure is comparable to the height of the tip.

The AFM is a powerful tool for imaging surfaces. With proper care, accurate images can be obtained and can provide important quantitative information about surfaces at the nanometer scale. However, it is imperative that researchers using this tool are aware of the many artifacts that can be present in the created images. These artifacts may be because of the shape and size of the AFM tip, the lateral or vertical interaction between the tip and the surface, or the electronics of the AFM system. Since the functionality of the AFM depends on the interaction between the tip and the surface, most of the artifacts that exist with the AFM are related to the size, shape, and cleanliness of the tip. Artifacts can also be introduced when the tip interacts with the surface in a way that changes the surface itself. Once aware of these artifacts, corrections can be made to create more accurate images of the surface structures of interest (Contescu 2009).

12.4 Electron Microscopy

Electron microscopy (EM) has long been used for ultrastructural analysis of both biological and non-biological samples. There are basically two types of electron microscopy, transmission and scanning electron microscopy (TEM and SEM). TEM can visualize internal subcellular structures from thin sliced cells, while SEM can yield more lifelike cell surface images. TEM is of high resolution, and it is one of the only few available instruments capable of resolving the structural features of nanoscale particles. When used in conjunction with detectors such as a backscattering detector (BSD) or energy dispersive X-ray spectroscopy (EDX) detector, SEM and TEM can be used to perform element analysis.

12.4.1 Transmission Electron Microscopy

Transmission electron microscopy (TEM) is one of the most frequently used tools for the characterization of nanomaterials. Aberration correction has revolutionized the field of electron microscopy, and now, instruments are commercially available providing sub-angstrom resolution and single-atom sensitivity for atomic, electronic, and chemical structure analyses (Contescu 2009).

In TEM, electrons emitted from a source are accelerated at high voltage potential and passed through a series of electromagnetic fields (conventionally called a "lens") (McNeil 2010). Some electrons pass through the thinly sliced (70–90 nm) TEM sections under study, while other electrons are scattered or diffracted by the sample. The electrons that pass through the sample move through another set of magnetic fields (called the objective, intermediate, and projection lens) and finally collide with a fluorescent screen. During this collision, their kinetic energy is converted to visible light energy, and this exposes a photographic film or excites a charge-coupled device (CCD) camera for digital imaging. This gives rise to a "shadow image" of the sample with different areas displayed with different darkness according to their density. Very thin slices of samples are required for TEM characterization so that the electrons can pass through the sample. Modern TEM has a resolution (the ability to distinguish two closely located points) of about 1 Å (angstrom, or 1×10^{-10} m). However, this does not mean one can always see biological molecules in TEM micrographs, since many biological molecules may not have a rigid structure or density capable of scattering high-velocity electrons. The electrons simply pass through the molecules and are therefore not visible in the resulting images. Techniques like embedding in epoxy resin (plastics) and flash freezing (cryo-TEM) can also be used to render some biological molecules sufficiently structured to scatter electrons and be visible in TEM micrographs. Figure 12.2 shows TEM pictures of colloidal silica.

12.4.2 Scanning Electron Microscope

The scanning electron microscope (SEM) is undoubtedly the most widely used of all electron beam instruments (Yao and Wang 2005). The popularity of the SEM can be attributed to many factors: the versatility of its various modes of imaging, the excellent spatial resolution now achievable, the very modest requirement on sample preparation and condition, the relatively straightforward interpretation of the acquired images, and the accessibility of associated spectroscopy and diffraction techniques. But most importantly, it is its user-friendliness, high levels of

Fig. 12.2 TEM pictures of colloidal silica

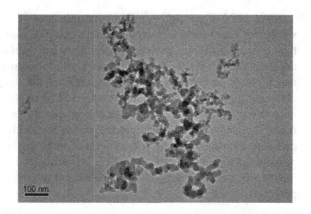

automation, and high throughput that makes it accessible to most research scientists.

When an electron beam interacts with a bulk specimen, a variety of electron, photon, phonon, and other signals can be generated. There are three types of electrons that can be emitted from the electron entrance surface of the specimen: secondary electrons with energies <50 eV, Auger electrons produced by the decay of the excited atoms, and backscattered electrons that have energies close to those of the incident electrons. All these signals can be used to form images or diffraction patterns of the specimen or can be analyzed to provide spectroscopic information. The de-excitation of atoms that are excited by the primary electrons also produces continuous and characteristic X-rays as well as visible light. These signals can be utilized to provide qualitative, semiquantitative, or quantitative information on the elements or phases present in the regions of interest. All these signals are the product of strong electron–specimen interactions, which depends on the energy of the incident electrons and the nature of the specimen.

Unlike in a TEM, where a stationary, parallel electron beam is used to form images, the SEM is similar to a fax/scanner machine or a scanning probe microscope and is a mapping device. In a SEM instrument, a fine electron probe, formed using a strong objective lens to demagnify a small electron source, is scanned over a specimen in a two-dimensional raster. Signals generated from the specimen are detected, amplified, and used to modulate the brightness of a second electron beam that is scanned synchronously with the SEM electron probe across a cathode-ray-tube (CRT) display. Therefore, a specimen image is mapped onto the CRT display for observation. If the area scanned on the sample is A_s and the corresponding area on the CRT display is A_d, then the magnification (M) of a SEM image is simply given by (Eq. 12.5):

$$M = \frac{A_d}{A_s} \qquad (12.5)$$

The SEM magnification is purely geometric in origin and can be easily changed by varying the scanned area on the sample.

Since SEM is a serial recording system instead of a parallel recording one, the whole process of generating a SEM image could be slower than that in the TEM. A high-quality SEM image usually builds up over several seconds to several minutes, depending on the types of signals; thus, high probe current within a small electron nanoprobe is desirable, and the microscope and the sample stability are critical in obtaining high-quality and high-resolution SEM images. Unlike in TEM, there is no rotation between the object and the image planes, and the microscope magnification can be changed without refocusing the electron beam so as to obtain an optimum focused image. The resolution of SEM images at high magnifications is primarily determined by the size of the incident electron probe, the stability of the microscope, and the sample and the inherent properties of the signal generation processes.

12.5 Energy Dispersive X-ray (EDX)

Though TEM can be employed to detect nanoparticles based on morphology, TEM alone cannot conclusively identify nanoparticles. Indeed, micrographs can be often ambiguous due to particle aggregation, contamination, or morphology change after cellular uptake. Energy dispersive X-ray (McNeil 2010) can be used to confirm the composition and distribution of the nanoparticles through spectrum and elemental mapping. Energy dispersive X-ray microanalysis is a technique used for identification of the elemental composition of a specimen. During EDX analysis, a specimen is bombarded with an electron beam inside a SEM. The bombarding electrons collide with the electrons of the specimen and displace them from their energy levels. A position vacated by an ejected inner shell electron is eventually occupied by a higher-energy electron from an outer shell. The electron transfer is accompanied by the release of energy through X-ray emission. The amount of energy released by the transferring electron depends on the energies of the initial and final shells. Atoms of each element release X-rays with unique amounts of energy during the transfer process. The "fingerprint" energies of the emitted X-rays can then be used to identify an element. Moreover, EDX microanalysis is capable of generating a map of one or more chemical elements of interest. This map is obtained by running the acquisition of X-ray spectra in scanning mode and letting the software determine the concentration of the element of interest at each point while imaging. The color can be coded in order to indicate the absolute or relative concentration of the element of interest, thus giving a 2D image of the abundance of a particular element. This map can be combined with transmission electron microscopy (TEM) or scanning electron microscopy (SEM) micrographs of the specimen in order to get information about the relative distribution of complementary or correlating elements. The spatial resolution of the elemental mapping is dependent upon various factors, including the accelerating voltage, beam concentration, detector limits, take-off angle, and noise-to-signal ratio. It is very important to develop a standardized technique that is capable of detecting the

Fig. 12.3 EDX pattern of colloidal silica

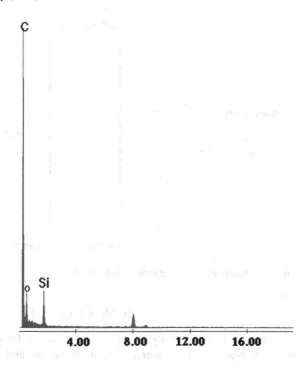

presence and distribution of nanoparticles in a tissue so as to confirm such properties as targeting and specificity. High-resolution TEM can be employed to detect nanoparticles based on their morphology. Figure 12.3 shows EDX pattern of colloidal silica.

12.6 Dissolution and Release Kinetics

The dissolution is the process by which solute molecules can be liberated from a solid phase and can enter into a solution phase. The dissolution begins with the initial detachment of solute molecules from the surface of the reservoir species. This reservoir depot could be drugs in their aggregated form or any delivery device designed to control the passage of solute into the solvent (known as the dissolution medium) (Babu et al. 2010).

The fundamental principle for evaluation of the kinetics of drug release was first established by Noyes and Whitney in (1897). The Noyes-Whitney equation gives the relation between the rate of dissolution of a dissolving solid in a given dissolution medium is directly proportional to the instantaneous concentration, (C) at time (t) and the saturation solubility (C_S) as given by Eq. 12.6 (Dokoumetzidis and Macheras 2006).

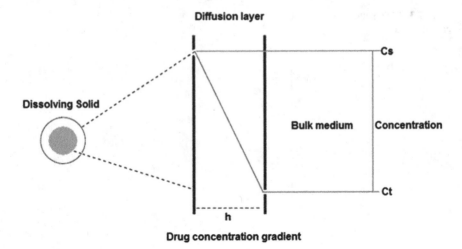

Fig. 12.4 Schematic representation of solid dissolution

$$dC/dt = k(C_s - C) \qquad (12.6)$$

The saturation solubility C_s of a drug is a key factor in the Noyes-Whitney equation. Nernst and Brunner further modified this and represented it as Eq. 12.7 (Dokoumetzidis and Macheras 2006).

$$dC/dt = DS(C_s - C_t)/Vh \qquad (12.7)$$

where dC/dt is the rate of dissolution of the solute, D is the diffusion coefficient of the solute in solution, S is the surface area of exposed solid, h is the thickness of the diffusion layer, and V is the volume of dissolution medium. $(C_s - C_t)$ is the driving force for the dissolution process (termed as concentration gradient driving force), where C_t is the concentration at time t (Fig. 12.4).

Dissolution of the solid is said to be occurring always under sink conditions (Dash et al. 2010), i.e., the quantity of the medium is all the time greater than the saturation solubility of the dissolving species (Fig. 12.4).

The purpose of any drug delivery system is to make available quickly a therapeutic amount of the drug to the proper site of the body and then maintaining the desired drug concentration throughout the therapy. The only way out for such problem is fabrication of a controlled-release drug delivery system which can be tailored to deliver the drug at a rate dictated by the needs of the body over a specified period of the treatment. The rationale for the controlled delivery of drugs is to promote the therapeutic benefits while at the same time minimizing the toxic effects. Controlled/sustained drug delivery can reduce the undesirable fluctuation of drug levels, enhancing therapeutic action and eliminating any dangerous side effects.

12.6.1 Goodness-of-Fit Model-Dependent Approach

The goodness-of-fit model-dependent approach can be employed to analyze the release kinetics (Thakkar et al. 2009 and Phaechamud et al. 2010). The in vitro data obtained from the dissolution experiments can be fitted to various kinetic models like zero order, first order, Higuchi, Hixson-Crowell cube root law, and Korsmeyer-Peppas model. The regression analysis can be performed for all the selected models. The regression coefficient value (r^2) obtained after linearization can be used to predict a mechanism of the drug release.

12.6.2 Zero-Order Model

The zero-order model is the release system in which the delivery rate of drug remains constant until the delivery device is exhausted of the active ingredient (Dash et al. 2010). Drug dissolution from dosage forms for the zero-order delivery model can be represented by Eq. 12.8:

$$Q_0 - Q_t = k_0 t \tag{12.8}$$

Rearrangement of Eq. 12.8 yields Eq. 12.9:

$$Q_t = Q_0 + k_0 t \tag{12.9}$$

where Q_t is the amount of drug dissolved in time t, Q_0 is the initial amount of drug in the solution (most times, Q_0 is zero), and k_0 is the zero-order release constant expressed in units of concentration/time. In the case of zero-order model graph, the data obtained from the in vitro drug release studies are plotted as percent cumulative amount of drug released versus time.

12.6.3 First-Order Model

The first-order model is used to describe absorption and/or elimination of some drugs (Dash et al. 2010 and Singh et al. 2011). The release rate in the first order is directly proportional to the amount of active ingredient load in the device.

The release of the drug which follows first-order kinetics can be expressed by Eq. 12.10:

$$dC/dt = -kC \tag{12.10}$$

where k is first-order rate constant expressed in units of time^{-1}.

Equation 12.10 can be expressed as:

$$\log C = \log C_0 - kt/2.303 \tag{12.11}$$

where C_0 is the initial concentration of drug, k is the first-order rate constant, and t is the time. The data obtained from the in vitro release studies are plotted as logarithmic cumulative percentage of drug remaining in the dosage form versus time.

12.6.4 Higuchi Model

Higuchi proposed the first example of a mathematical model that aimed to describe drug release from a matrix system in 1961 (Dash et al. 2010 and Singh et al. 2011). This model is based on the hypotheses that (1) initial drug concentration in the matrix is much higher than drug solubility, (2) drug diffusion takes place only in one dimension (edge effect must be negligible), (3) drug particles are much smaller than system thickness, (4) matrix swelling and dissolution are negligible, (5) drug diffusivity is constant, and (6) perfect sink conditions are always attained in the release environment. Accordingly, the model expression is given by the Eq. 12.12

$$Q = A\sqrt{D\,(2C - C_s)Cs\,t} \tag{12.12}$$

where Q is the amount of drug released in time t per unit area A, C is the initial drug concentration, C_s is the drug solubility in the matrix media, and D is the diffusivity of the drug molecules (diffusion coefficient) in the matrix substance. This relation is valid all the time, except when the total depletion of the drug in the therapeutic system is achieved. To study the dissolution from a planar heterogeneous matrix system, where the drug concentration in the matrix is lower than its solubility and the release occurs through pores in the matrix, the expression is given by Eq. 12.13.

$$Q = \sqrt{D(\delta/\tau)(2C - \delta C_s)Cs\,t} \tag{12.13}$$

where D is the diffusion coefficient of the drug molecule in the solvent, δ is the porosity of the matrix, τ is the tortuisity of the matrix, and $Q, A, C_s,$ and t have the meaning assigned above. Tortuisity is defined as the dimensions of radius and branching of the pores and canals in the matrix. In a general way, it is possible to simplify the Higuchi model (Eq. 12.13) as (generally known as the simplified Higuchi model)

$$Q = k_H \times t^{1/2} \tag{12.14}$$

where k_H is the Higuchi dissolution constant.

The data obtained from the in vitro release kinetic studies can be plotted as cumulative percentage drug release versus square root of time.

12.6.5 Hixson-Crowell model

Hixson and Crowell recognized that the regular area of a particle is proportional to the cube root of its volume (Dash et al. 2010 and Singh et al. 2011). They derived Eq. 12.15 as

$$W_0^{1/3} - W_t^{1/3} = \kappa t \qquad (12.15)$$

where W_0 is the initial amount of drug in the pharmaceutical dosage form, W_t is the remaining amount of drug in the pharmaceutical dosage form at time t, and κ (kappa) is a constant incorporating the surface–volume relation. The equation describes the release of drug from systems where there is a change in surface area and diameter of matrix system. To study the release kinetics, data obtained from in vitro drug release studies can be plotted as cube root of drug percentage remaining in the matrix versus time.

12.6.6 Korsmeyer–Peppas Model for Mechanism of Drug Release

Korsmeyer et al derived an equation that described drug release from a polymeric system (Dash et al. 2010). In order to find out the mechanism of drug release, an initial 60 % drug release data can be fitted in Korsmeyer–Peppas model (Eq. 12.16)

$$M_t/M_\infty = kt^n \qquad (12.16)$$

where M_t/M_∞ is fraction of drug released at time t, k is the rate constant, and n is the release exponent. The n value is used to characterize different release mechanisms as given in Table 12.1.

To study the release kinetics, data obtained from in vitro drug release studies can be plotted as log cumulative percentage drug release vs. log time (Singh et al. 2011).

Table 12.1 Interpretation of diffusional release mechanisms for Korsmeyer–Peppas model (Costa and Sousa Lobo 2001)	Diffusion exponent (n)	Drug transport mechanism
	0.5	Fickian diffusion
	$0.5 < n < 1.0$	Anomalous (non-Fickian) diffusion
	1.0	Case-II transport
	$n > 1.0$	Super-case-II transport

References

Babu VR, Areefulla S, Mallikarjun V (2010) Solubility and dissolution enhancement: an overview. J Pharm Res 3(1):141–145

Contescu CI (2009) Dekker encyclopedia of nanoscience and nanotechnology. CRC Press, Boca Raton

Costa P, Sousa Lobo JM (2001) Modeling and comparison of dissolution profiles. Eur J Pharm Sci 13(2):123–133

Dash S, Murthy PN, Nath L, Chowdhury P (2010) Kinetic modeling on drug release from controlled drug delivery systems. Acta Pol Pharm 67(3):217–223

Dokoumetzidis A, Macheras P (2006) A century of dissolution research: from Noyes and Whitney to the biopharmaceutics classification system. Int J Pharm 321(1):1–11

McNeil SE (2010) Characterization of nanoparticles intended for drug delivery. Humana Press, Clifton

Noyes AA, Whitney WR (1897) The rate of solution of solid substances in their own solutions. J Am Chem Soc 19(12):930–934

Phaechamud T, Mueannoom W, Tuntarawongsa S, Chitrattha S (2010) Preparation of coated valproic acid and sodium valproate sustained-release matrix tablets. Indian J Pharm Sci 72(2):173

Singh J, Gupta S, Kaur H (2011) Prediction of in vitro drug release mechanisms from extended release matrix tablets using SSR/R2 Technique. Trend Appl Sci Res 6:400–409

Thakkar V, Shah P, Soni T, Parmar M, Gohel M, Gandhi T (2009) Goodness-of-fit model-dependent approach for release kinetics of levofloxacin hemihydrates floating tablet. Dissolution Technol 16:35–39

Yao N, Wang ZL (2005) Handbook of microscopy for nanotechnology. Kluwer Academic Publishers, Dordrecht

About the Authors

Prof. Subho Mozumdar, PhD (State University of New York at Buffalo), is currently Associate Professor of Chemistry at the University of Delhi. Widely heralded today as a leading figure in Indian nanotechnology, Prof. Mozumdar returned to India in 1998 after postdoctoral work with Prof. Larry Grossman at Johns Hopkins University. His work has resulted in multiple patents that have been bought by leading companies. He has published in leading journals in the field. In recognition of his discoveries, he recently became the Academic Editor of PLOS ONE.

Arnab De, M.A, M.Phil, is currently a PhD candidate at Columbia University Medical Center. Before this, he was at Indiana University, Bloomington, where he worked with Prof. Richard DiMarchi (Standiford H. Cox Professor of Chemistry and the Linda & Jack Gill Chair in Biomolecular Sciences) to develop peptide-based prodrugs as therapeutics for diabetes. The work with Prof. DiMarchi resulted in two patents (licensed by Marcadia Biotech, recently acquired by Roche) and multiple publications in peer-reviewed journals. He presented his findings in the American Peptide Symposium 2009 and received the Young Investigator's Award. He subsequently came to Columbia University where he is developing transgenic mice to serve as potential models for autoimmune diseases.

Rituparna Bose, PhD (Indiana University, Bloomington), is currently an adjunct Assistant Professor at the City University of New York and has been interviewed as an expert in the field of biodiversity by the Times of India and Statesman, Calcutta (leading news daily in India). She serves in the Editorial Boards of multiple well-known journals and is an Editor of Acta Palaeontologica Sinica (published by the Chinese Academy of Sciences) and the Associate Editor-in-Chief of the International Journal of Environmental Protection.

Ajeet Kumar was a graduate student with Prof. Subho Mozumdar at the University of Delhi, India.

A. De et al., *Targeted Delivery of Pesticides Using Biodegradable Polymeric Nanoparticles*, SpringerBriefs in Molecular Science, DOI: 10.1007/978-81-322-1689-6, © The Author(s) 2014